Information Systems Engineering and Management 48

The book series "Information Systems Engineering and Management" (ISEM) publishes innovative and original works in the various areas of planning, development, implementation, and management of information systems and technologies by enterprises, citizens, and society for the improvement of the socio-economic environment.

The series is multidisciplinary, focusing on technological, organizational, and social domains of information systems engineering and management. Manuscripts published in this book series focus on relevant problems and research in the planning, analysis, design, implementation, exploration, and management of all types of information systems and technologies. The series contains monographs, lecture notes, edited volumes, pedagogical and technical books as well as proceedings volumes.

Some topics/keywords to be considered in the ISEM book series are, but not limited to: Information Systems Planning; Information Systems Development; Exploration of Information Systems; Management of Information Systems; Blockchain Technology; Cloud Computing; Artificial Intelligence (AI) and Machine Learning; Big Data Analytics; Multimedia Systems; Computer Networks, Mobility and Pervasive Systems; IT Security, Ethics and Privacy; Cybersecurity; Digital Platforms and Services; Requirements Engineering; Software Engineering; Process and Knowledge Engineering; Security and Privacy Engineering, Autonomous Robotics; Human-Computer Interaction; Marketing and Information; Tourism and Information; Finance and Value; Decisions and Risk; Innovation and Projects; Strategy and People.

Indexed by Google Scholar. All books published in the series are submitted for consideration in the Web of Science.

For book or proceedings proposals please contact Alvaro Rocha (amrrocha@gmail.com).

Franci Suni-Lopez · Elvira G. Rincón Flores ·
Hernan Alejandro Quintana Cruz ·
Eunice Pereira dos Santos Nunes
Editors

Extended Reality and Serious Games for Education, Competitiveness, and Wellbeing

Proceedings of CIIS 2024

Springer

Editors
Franci Suni-Lopez
Universidad de Lima
Santiago de Surco, Peru

Hernan Alejandro Quintana Cruz
Universidad de Lima
Santiago de Surco, Peru

Elvira G. Rincón Flores
Tecnológico de Monterrey
Monterrey, Nuevo León, Mexico

Eunice Pereira dos Santos Nunes
Universidade Federal de Mato Grosso
Cuiabá, Mato Grosso, Brazil

ISSN 3004-958X ISSN 3004-9598 (electronic)
Information Systems Engineering and Management
ISBN 978-3-031-91480-5 ISBN 978-3-031-91481-2 (eBook)
https://doi.org/10.1007/978-3-031-91481-2

This Springer imprint is published by the registered company Springer Nature Switzerland AG
The registered company address is: Gewerbestrasse 11, 6330 Cham, Switzerland

Preface

The CIIS 2024 (International Conference of Systems Engineering) is an annual conference organized by the systems engineering program at the Universidad de Lima, in Peru. It serves as a platform for discussions on various areas of computing and systems engineering, positioning itself as a regional forum for cutting-edge research.

The 2024 conference marks the 7th edition of the event, reaffirming its commitment to fostering collaboration and innovation. With a record number of submissions this year, the program reflects the diversity and depth of research being conducted in different computing areas. The selected papers, covering topics such as AI and data analytics, software engineering, information technology, and the social and cultural impact of technology, represent cutting-edge contributions that push the boundaries of our understanding and capabilities.

The conference featured five keynote presentations addressing the main theme of the congress: the development of video games and virtual reality applications. These were followed by the presentation of papers from various fields of computing and systems engineering.

In the 2024 edition, we received 51 submissions, of which 15 were accepted, resulting in a 29% acceptance rate. The article selection process involved a double-blind peer review by reviewers with proven expertise in the field. Each paper was assigned two reviewers, and in cases where there was no agreement between them, a third reviewer was called in to decide on the article's acceptance.

We extend our gratitude to the authors for their high-quality submissions and to the program committee for their rigorous reviews. Their efforts ensure the proceedings maintain the highest academic standards. Special thanks also goes to the keynote speakers, whose thought-provoking talks set the tone for stimulating discussions throughout the event.

Enjoy the proceedings and the insights they offer.

December 2024

Franci Suni-Lopez
Hernan Alejandro Quintana Cruz
Elvira G. Rincón Flores
Eunice Pereira dos Santos Nunes

Organization

CIIS 2024 is organized by the systems engineering program, at the Universidad de Lima, in Lima, Peru.

Organizing Committee

Conference Chair

Franci Suni-Lopez	Universidad de Lima, Peru

Conference Co-chair

Elvira G. Rincón Flores	Tecnológico de Monterrey, Mexico

Program Chair

Hernan Quintana-Cruz	Universidad de Lima, Peru

Program Co-chair

Eunice Pereira dos Santos Nunes	Universidade Federal de Mato Grosso, Brazil

Scientific Committee

Cristiano Maciel	Universidade Federal de Mato Grosso, Brazil
Edwin Escobedo Cárdenas	Universidad de Lima, Peru
Effie Lai-Chong Law	Durham University, England
Guillermo Antonio Dávila Calle	Universidad de Lima, Peru
Hernán Nina Hanco	Universidad de Lima, Peru
Isaias Bianchi	Federal University of Santa Catarina, Brazil
Manuel Castillo Cara	Universidad de Castilla-La Mancha, Spain
Marco Antonio Sotelo Monge	Indra, Spain
Michael Dorin	University of St. Thomas, USA
Nelly Condori Fernandez	Universidad Santiago de Compostela, Spain
Ruth Maria Reategui Rojas	Universidad Técnica Particular de Loja, Ecuador

Organizing Committee

Local Organizing Committee

Nadia Rodriguez Rodriguez	Universidad de Lima, Peru
Andrea Matuk Chijner	Universidad de Lima, Peru
Jim Bryan Dios Luna	Universidad de Lima, Peru
Carlos Torres Paredes	Universidad de Lima, Peru
Pablo Rojas Jaen	Universidad de Lima, Peru
Rosario Guzman Jiménez	Universidad de Lima, Peru
Karla Alejandra Gomez Loayza	Universidad de Lima, Peru
Javier More Sanchez	Universidad de Lima, Peru

Referees

Josmell Alva	Universidad de Trujillo, Peru
Edwin Alvarez	Universidad de Lima, Peru
Renzo Angles	Universidad de Talca, Chile
Victor Ayma	Universidad Pacífico, Peru
Luis Barba	Universidad Técnica Particular de Loja, Ecuador
Jose Caballero	Universidad de Lima, Peru
Fiorella Capcha	Universidad de Lima, Peru
Ricardo Chavez	Abo Akademi, Finland
Josimar Chire	University of São Paulo, Brazil
Ana Cuadros	Universidad Nacional San Agustin, Peru
Guillermo Antonio Davila Calle	Universidad de Lima, Peru
Bertha Diaz	Universidad de Lima, Peru
Ignacio Diaz-Cano	Universidad de Cadiz, España
Jim Dios	Universidad de Lima, Peru
Edwin Escobedo	Universidad de Lima, Peru
Jorge Fernandez	Computer Science Research Institute of Toulouse, France
Roxana Flores	Universidad Nacional San Agustin, Peru
Miguel Gonzalez Mendoza	Tecnológico de Monterrey, Mexico
Juan Gutierrez	Universidad de Lima, Peru
Rosario Guzman	Universidad de Lima, Peru
Edgar Jesus Huaranga Junco	Universidad de Lima, Peru
Jorge Irey	Universidad de Lima, Peru
Valdivia Jose	Universidad de Lima, Peru
Cynthia López	Tecnológico de Monterrey, Mexico

Vicente Machaca	Universidad La Salle, Peru
Paulo Maia	Universidad de Lima, Peru
Edgar Manzano	Universidad de Trujillo, Peru
Andrea Matuk Chijner	Universidad de Lima, Peru
Angela Mayhua	Universidad de Lima, Peru
Markus Mock	Landshut University of Applied Sciences, Alemania
Javier More	Universidad de Lima, Peru
Hernan Nina	Universidad de Lima, Peru
Wilder Nina Choquehuayta	Universidad de Ingeniería y Tecnología, Peru
Brenner Ojeda	Universidad de Ingeniería y Tecnología, Peru
Rosa Paccotacya	Universidad Nacional San Agustin, Peru
Luis Paredes	Pontificia Universidad Católica del Peru, Peru
Hernan Quintana-Cruz	Universidad de Lima, Peru
Lennin Paul Quiroz	Universidad de Lima, Peru
Marcela Quispe	Universidad Católica San Pablo, Peru
Lourdes Ramirez	Universidad de Lima, Peru
Elvira Rincon	Tecnológico de Monterrey, Mexico
Guillermo Rodriguez	Universidad Nacional del Centro de la Pcia. de Buenos Aires, Argentina
Nadia Katherine Rodriguez Rodriguez	Universidad de Lima, Peru
Pablo Rojas	Universidad de Lima, Peru
Donna Roper	Universidad Pacífico, Peru
Brenda Santos	Tecnológico de Monterrey, Mexico
Edgar Sarmiento	Universidad Nacional San Agustin, Peru
Andrew Seely	University of Maryland Global Campus, USA
Ernesto Suarez	Universidad Nacional San Agustin, Peru
Franci Suni	Universidad de Lima, Peru
Dayannis Tamayo	Tecnológico de Monterrey, Mexico
Fernando Teshima	Universidad de Lima, Peru
Regina Ticona	Universidad Católica San Pablo, Peru
Carlos Torres	Universidad de Lima, Peru
Yvan Tupac	Universidad Católica San Pablo, Peru
Jorge Veloz	Universidad Técnica de Manabí, Ecuador
Nadia Yoza	University of Colorado Boulder, USA
Guillermo Zevallos	Universidad de Lima, Peru

Contents

Videogame Applications

Urarina Field Note Restoration Using Machine Learning and Game Play

Zhengyuan Feng and Michael Dorin(✉)

University of St. Thomas, St. Paul, MN, USA
{feng2052,mike.dorin}@stthomas.edu
http://www.stthomas.edu

Abstract. Before the common availability of portable electronics, researching endangered languages required recording voices and writing lexicon cards. Lexicon cards describe a word and provide phonetic symbols depicting its pronunciation. In a typical study, multiple researchers with different handwriting styles may produce these cards. Variety in writing styles and the assortment of symbols used often makes optical character recognition difficult. This research addresses data capture challenges with a multi-phase process for accurately digitizing handwritten lexicon cards. First, lexicon cards were scanned into images and submitted to Google Cloud Vision for processing. Google Cloud Vision returned the recognized characters and mathematical bounding boxes denoting the physical locations of all text on the cards. Next, deep learning was employed to decode the phonetic symbols. These symbols were extracted manually using the bounding boxes provided by Google. A convolutional neural network-based application then processed the images and stored the most promising prediction of the symbol that matched the image. Because the automated processes thus far were not 100 % accurate, a further step involved human review and editing. Manual editing is commonly accepted as tedious and error-prone, indicating it also did not meet accuracy goals. An essential final step was creating a game to encourage review of the digital results. Not only does the game encourage an additional review, but it simultaneously provides practice and training to linguists studying the language. Through this process, the digitization of lexicon cards reached 100 % accuracy. This new approach can significantly help revitalize dormant language studies.

Keywords: Google Cloud Vision · Optical Character Recognition · Games · Urarina Language

---

Supported by The University of St. Thomas.

F. Suni-Lopez et al. (Eds.): CIIS 2024, ISEM 48, pp. 3–17, 2025.
https://doi.org/10.1007/978-3-031-91481-2_1

1 Introduction

There are many tools available to the modern researcher, such as the availability of powerful digital computers. Nonetheless, many valuable research projects were conducted before the advent of accessible computers. For example, a Google Scholar search for linguistics projects between 1900 and 1960 yielded about 16,700 results [31]. Historical projects can provide useful data and information to inspire contemporary research. A prominent obstacle faced by those wishing to revitalize classic projects is data is often stored in paper form, sometimes including handwritten data.

The inspiration for this project came from the efforts of a group of students from the University of San Marcos who, in 1987, embarked on a project to study the language spoken by the Urarina people. The Urarina people, who reside in the Chambira River Basin, number only two to three thousand people. According to the United Nations Educational, Scientific and Cultural Organization (UNESCO), their language is "definitely endangered. [37]" This project's work is not only crucial for the restoration of the Urarina study but also holds promise for other language research projects, potentially helping to resolve the trend of language extinction.

In the Urarina study, lexicon cards were created to record and manage vocabulary items from the Urarina language [1]. The lexicon cards contain information about words, including pronunciation, definition, and usage data. Because portable computers were unavailable, these cards were all handwritten. An example lexicon card can be seen in Fig. 2. These cards are an important part of linguistic field notes and have similarities to dictionary cards [22]. This project studies the application of modern technology to restore and digitally preserve this valuable data. This project is important because the concepts discovered can be applied to many other historical works.

Several tools already exist, with their capability to perform similar tasks being advertised. For example, Google provides cloud applications specifically to perform optical character recognition (OCR) through the Google Cloud Vision API [27]. Amazon and Microsoft provide a similar cloud-based service, with published interfaces allowing custom applications to use them [23,32]. Numerous other offerings exist, including open-source solutions such as Tesseract OCR [35]. Although these services are powerful and useful, their implementation is designed to serve more generic character recognition tasks, and they are often difficult to apply to custom symbols. As such, these tools are beneficial for certain aspects of this project but not sufficient for the entirety of the work.

Along with these common off-the-shelf tools, numerous research studies have been carried out on custom OCR and symbol identification systems. Many researchers have explored convolutional neural networks (CNN). For example, in their paper, "CNN based Optical Character Recognition and Applications," Sarika et al. evaluate various algorithms for using a CNN to perform character recognition [30]. They show good results and have an excellent section describing applications of their work. Narayan and Muthalagu also use CNNs. They describe how to use them for unique and handwritten characters [25]. They give a good explanation of their system and show good results. A recent paper by Yu et al. shows how to use a contrastive language image pre-training model for scene detection [39]. A paper by Mamun et al. even covers using a CNN for the classification of non-native handwritten characters [19]. This is an excellent paper that provides valuable insight for the current project. Analogous to the previously mentioned tools, the aforementioned research works benefit specific aspects of this research, but by themselves, do not accomplish the restoration and digitization goals of this project. Games have been previously employed in linguistic research. Works, such as those by Matyas et al. [21] and Saarenpää [29], show data-gathering techniques using games. However, the data gathered is more focused on collecting new behavioral data rather than validating what may be presented in a game. Past studies have shown that linguistics games aid knowledge acquisition [17,26,40]. However, these studies focus on learning rather than data validation.

Because applying existing works to the restoration process with sufficient accuracy is impossible, a different and novel approach has been created for this language research. In the first step, images of the cards were submitted to Google Vision for evaluation and OCR operations. The results provided by Google give a reasonable estimate of text and bounding boxes of all text [27]. Next, the special characters from the lexicon cards-manually extracted or extracted by Google into bounding boxes-are processed by a deep learning model. Upon completion, a human examination was still necessary to confirm the accuracy of the capture. Human review is a mundane task that can be laborious and susceptible to errors. Although automation strives to eliminate this type of work [2,38], complete automation is impossible for this project. So, this paper presents the innovative concept of using a game as a vital part of the restoration process. Through the game, the results of the digitization are verified, and the player receives valuable practice in the art of linguistics. This essay shows how to use state-of-the-art tools combined with human interaction to accurately digitize lexicon cards from historic project.

2 Methods

The innovative approach to the digital restoration of decades-old lexicon cards encompasses different steps. A summary is illustrated in the *Card Restoration* data flow diagram (DFD), Fig. 1. In this study, lexicon cards typically contained two types of information. First is the phonetic spelling of the word's pronunciation. The special characters in the phonetic spelling represent the phones-the basic speech sounds used when the word is spoken [18,36]. Additional information provided on the card includes an equivalent Spanish word and normally a description, in Spanish, of what was said by the Urarina speaker.

2.1 Scanning Lexicon Cards

The DFD shows *Scanning Cards* as Process #1 (Fig. 1). The handwritten lexicon cards were manually scanned using a Fujitsu card scanner [9] to build the *Card*

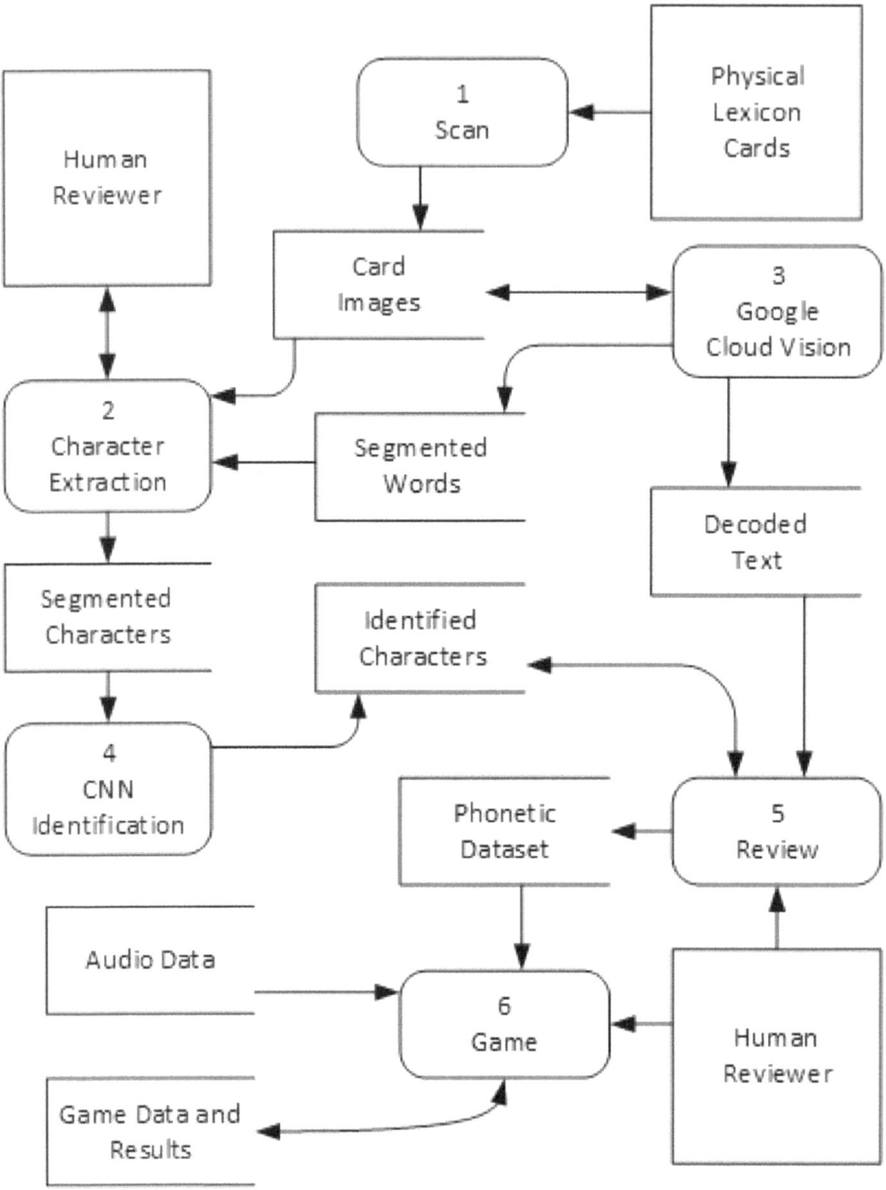

Fig. 1. Dataflow Diagram of the Card Restoration Methodology.

Images data store. Each symbol was organized into folders named after the symbol itself, with images sequentially numbered to facilitate easy access and management.

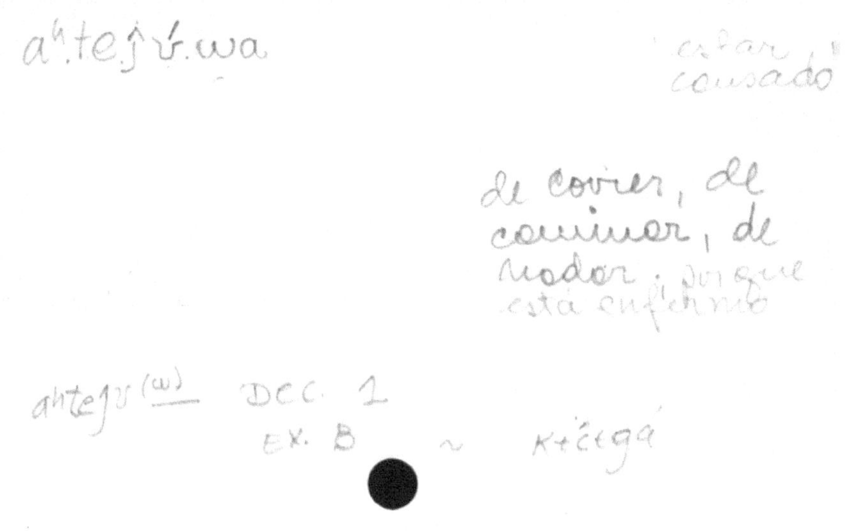

Fig. 2. Sample Lexicon Card.

2.2 Character Extraction and Google Cloud

Once the cards were scanned, individual character images were extracted for two purposes. The first purpose was to create test and validation data for the objective of creating a CNN character identification application. Character extraction was done both manually using an image editing application and with some automated support through Google-provided bounding boxes.

Manual Extraction. To build the dataset of phonetic symbols for the Urarina language, character symbols were cropped from the images of the handwritten card. For an example, see Fig. 3. Symbols were stored in the *Segmented Characters* data store. This data store was constructed using folders named after the symbol itself, with images numbered sequentially. To preserve image integrity, all images were saved in bitmap (BMP) format, which is known for its reliable storage of visual data without loss [12]. The project approach ensured a systematic collection of diverse symbols, including both commonly used symbols from lexicon cards and those extracted from the lexicon cards, as well as additional handwritten sources. Manual extraction was only used to create datasets to train the CNN.

Google Cloud Vision API. was used to assist in the automation of character extraction. Cloud Vision has many features, but two important ones for this project were advanced OCR and bounding box identification. A Python application was written to submit lexicon card images to the Google Cloud Vision

Fig. 3. Phonetic Symbol of retroflex (left) and bilabial nasel (right).

service. A *Decoded Text* data store was created for the Spanish information found on the card. A *Product Characters* data store was created for the word segmentation information. This data store maintains pointers to the original lexicon card from which the characters originated.

2.3 Deep Learning, The Convolutional Neural Network

The DFD shows *Deep Learning* as Process #4 (Fig. 1). To ensure a sufficient dataset size, each symbol was represented by 80 to 110 images [28]. This range was selected to ensure an accurate and robust symbol recognition system by providing ample training and testing data. In total, 86 individual symbols were collected for this project. Figure 6 shows an example of each image.

The raw images collected from the handwritten cards were preprocessed to create the training dataset for analysis. Each image was resized to a standardized format of 40×40 pixels and converted to grayscale color mode to ensure uniformity in representation. Following preprocessing, the dataset was split into training and testing subsets using a 90/10 ratio. This systematic approach to data preparation ensured consistency in input dimensions and facilitated effective evaluation of the CNN model's performance [5].

This CNN is structured to handle the variability of lexicon card handwriting styles. For feature extraction, the architecture begins with a 2D convolutional layer (Conv2D) with 64 filters and a kernel size of 3×3. This will capture spatial hierarchies like edges and textures necessary for distinguishing different handwritten symbols. The choice of 64 is a balance between model complexity and computational efficiency [14,24].

A batch normalization layer (BatchNormalization) is applied to normalize the output from the convolutional layer. This step reduces internal covariate shift, speeds up convergence, and helps mitigate overfitting by allowing the network to use higher learning rates [13]. The next layer is the Rectified Linear Unit (ReLU) activation function. This introduces non-linearity into the model, which is helpful for learning complex patterns. It was decided to use ReLU instead of sigmoid because it is less prone to the vanishing gradient problem, where gradients become very small during backpropagation, slowing down or halting learning. [10,33].

After the activation function, a MaxPooling2D layer with a pool size of 3×3 and a stride of 2×2 is added. This is to reduce the spatial dimensions of the feature maps. Max pooling helps retain the most prominent features while discarding less relevant information, further reducing the computational load and controlling overfitting [16].

This combination of convolution, batch normalization, activation, and pooling is repeated multiple times to allow the network to learn progressively higher-level features [8]. Now a Flatten layer is used to convert the 2D feature maps into a 1D vector. This transformation is necessary to move from the convolutional part of the network to the fully connected (dense) layers, which operate on 1D inputs. A dense layer with 64 neurons and ReLU activation is then employed to learn complex combinations of the extracted features [15,22]. The output layer is a dense layer with ten neurons (numClasses = 10), one for each symbol class, using the softmax activation function to convert the outputs into a probability distribution [7].

This implementation comprising convolutional, normalization, activation, pooling, and dense layers, was designed to effectively learn and classify phonetic symbols from images prepared during data preprocessing. This can be seen in Fig. 4. It allows the network to predict the most likely class for each input image. The model is compiled using the Stochastic Gradient Descent optimizer with a learning rate of 0.001 [3]. This learning rate was chosen for its effectiveness in large-scale datasets and its ability to provide stable convergence [34]. Finally, a categorical cross-entropy loss function is employed, which is appropriate for multi-class classification tasks. Training is conducted separately for each group of symbols with an epoch range of 50 to 75 and a batch size of 35, providing a good balance between learning time and model performance [20].

The dataset includes 86 different phonetic symbols representing phones from the Urarina language. These are shown in Fig. 6. However, due to the rarity of some letters, this project focused on the 36 most important symbols, grouping the remaining ones into a category called "everything else." For training the CNN, symbols were organized into groups of ten, including the "everything else" group. This batching strategy helps the model learn more effectively by focusing on smaller sets of related symbols [6]. The project trained four different CNN models to assess the impact of various training settings on performance. Although the CNN layers remained consistent across all models, the learning rate, the number of epochs, and the mini-batch size varied.

You can see the confusion matrix in Fig. 7, which shows the results from Group 1. The symbols in this group are: "K": 0, "a": 1, "a-aspirated": 2, "a-nasalized": 3, "e-aspirated": 4, "h": 5, "everything else": 6, "retroflex": 7, "voiceless-fricative-labialized": 8, and "voiceless-plosive-palatalized": 9. During testing, CNN achieves an average accuracy of about 85 %. This high accuracy demonstrates CNN's ability to classify symbols accurately despite the large variety of classes. Figure 5 shows the training validation accuracy and loss.

After training, the project had four individual CNNs created from batches of nine phonetic symbols from the dataset. Each CNN represents a subset of the total 37 classes (36 important symbols plus an additional category for miscellaneous symbols named "everything else"). With these CNNs, work on the project proceeded to classify the test images. For each test image, test symbols were provided to all four trained CNN models to obtain predictions from each group. It was then possible to select the predicted class as the final prediction.

```
Model: "sequential"
_________________________________________________________________
 Layer (type)                Output Shape              Param #
=================================================================
 conv2d (Conv2D)             (None, None, None, 64)    1792

 batch_normalization (Batch  (None, None, None, 64)    256
 Normalization)

 activation (Activation)     (None, None, None, 64)    0

 max_pooling2d (MaxPooling2  (None, None, None, 64)    0
 D)

 flatten (Flatten)           (None, None)              0

 dense (Dense)               (None, 64)                1327168

 dense_1 (Dense)             (None, 9)                 585

=================================================================
Total params: 1329801 (5.07 MB)
Trainable params: 1329673 (5.07 MB)
Non-trainable params: 128 (512.00 Byte)
_________________________________________________________________
```

Fig. 4. CNN Layers

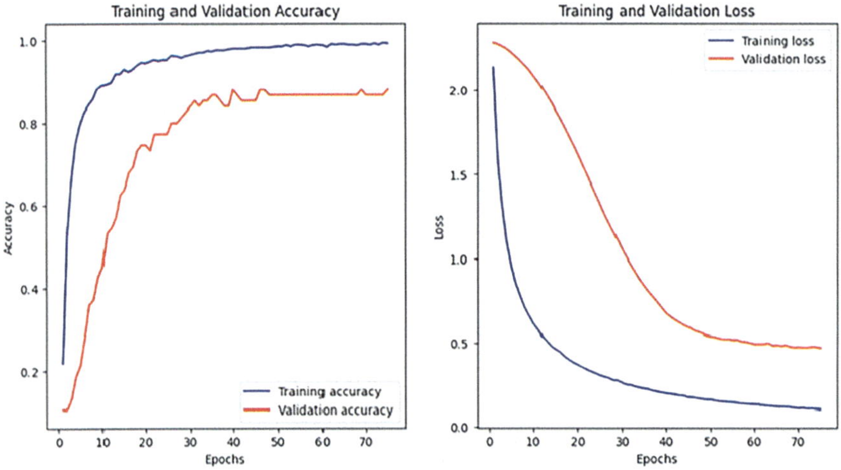

Fig. 5. Training and Validation Accuracy

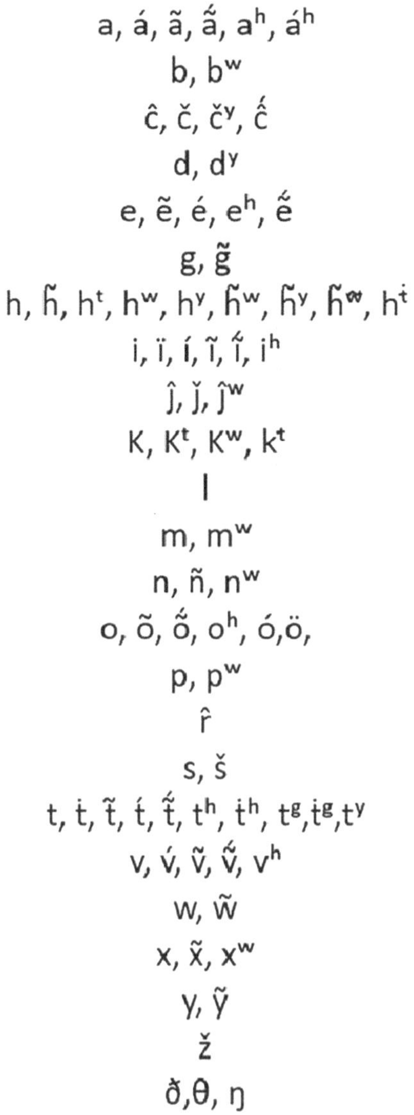

Fig. 6. Individual Phonetic Symbols

Ideally, if a symbol is selected from one of the 36 important symbols, the four CNNs should result in three predicting the symbol as "everything else" and one predicting the true class. If a rare symbol is chosen, all four CNNs should put it in the "everything else" class. However, due to the confusion matrix not achieving 100% accuracy and rare symbols not specified in the CNN, the final step involves collecting all four prediction results and performing a manual check to

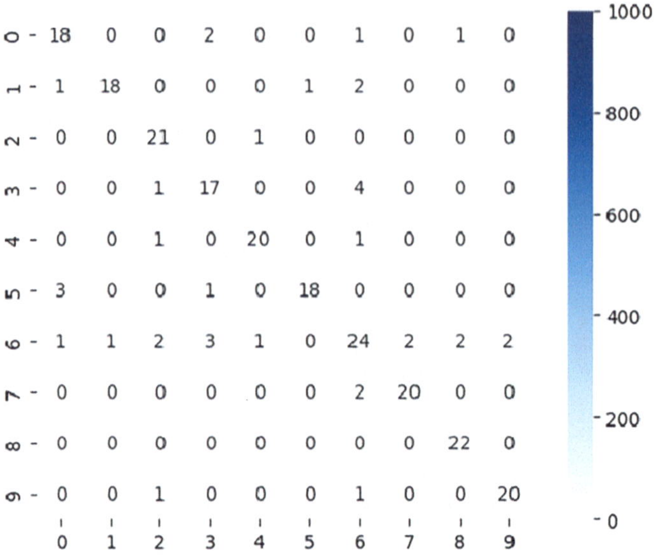

Fig. 7. Confusion Matrix for Group 1 Test Dataset

Group	Symbols (w/o everything else)	Learning rate	Epoch	Batch size	Accuracy
1	a,h,K, a^h, e^h, ã, t^y, p^w, ĵ	0.002	60	50	88.3%
2	č,n,x,s,o,t,e,v é	0.001	80	35	87.6%
3	m,i,d, ñ, t̃, ṽ́, ẽ, ĩ, á	0.001	75	35	83.6%
4	b,g,l,y, í, i^h, ỹ, v́, ṫ	0.002	70	35	85.4%

Fig. 8. CNN Inputs, Parameters and Accuracy

ensure accuracy. Figure 8 summarizes various inputs, chosen parameters, and final accuracies.

Figure 9 shows the "i-asperiated i^h" symbol that required a manual review. The predictions from the four CNNs came back as "a-asperiated, everything else, everything else, and i-asperiated." Thus, a manual review was later required for proper identification of this symbol as "i-asperiated."

2.4 Data Review

The DFD shows the *Data Review* as Process #5 (Fig. 1). Once the deep learning application finished classifying phonetics characters, the results were combined

Fig. 9. Hand drawn i-asperiated i^h

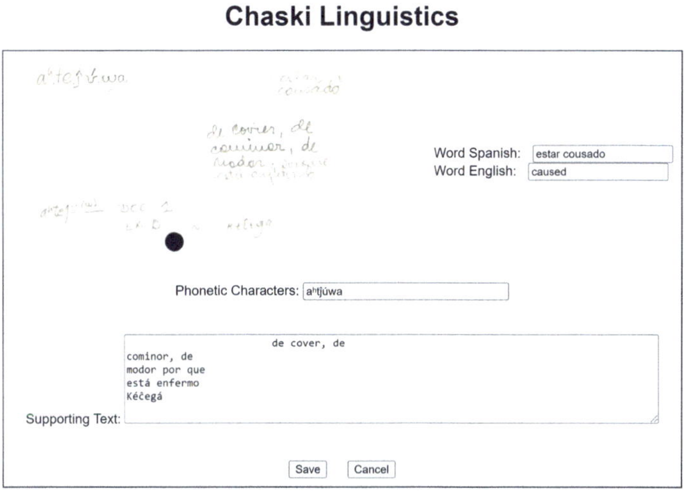

Fig. 10. Edit Application for Human Review

with the supporting Spanish text for a final human review. (Remember, Google Cloud Vision recognized the Spanish text separately.) An application displays the original card, the characters it believes it has classified, and the Spanish text provided by Google Cloud Vision. This web-based application allows editing of all fields, including Spanish and phonetic characters. See Fig. 10 for a screenshot of the application.

2.5 The Game

The DFD shows the *Game* as Process #6 (Fig. 1). Although the game portion of this project was initially written to cope with the tedium of editing, it quickly became apparent that it was a crucial aspect of the project. The web-based game has two modes. Mode one allows the player to listen to an Urarina word and then choose a phonetic spelling from a provided list. This simple activity allows for the practice of sound identification training and requires a reading of the lexicon card. The user interface for this mode is shown in Fig. 11.

A more challenging aspect of the game displays a lexicon card with some random phonetic characters missing, requiring the player to identify the missing phonetic characters. An indispensable feature of this game is that the system

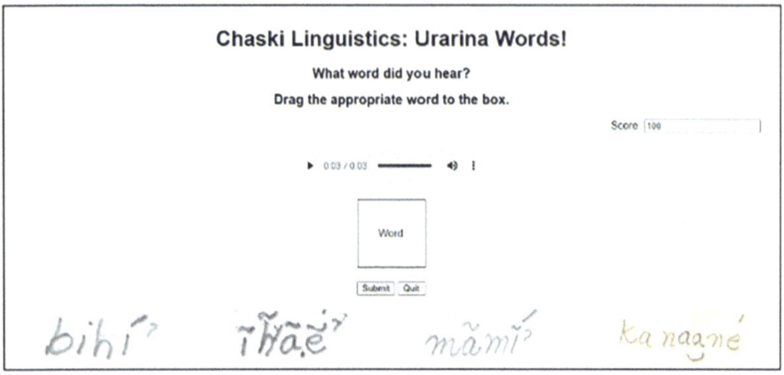

Fig. 11. Linguistics Word Game User Interface

tracks mistaken identification attempts. The card is marked for a fresh review for common errors since it is possible the original writer of the card may have made a mistake. Once marked for review, an expert can review the lexicon card, and the data can be updated as necessary. A simplified user interface for this part of the game is illustrated in Fig. 12.

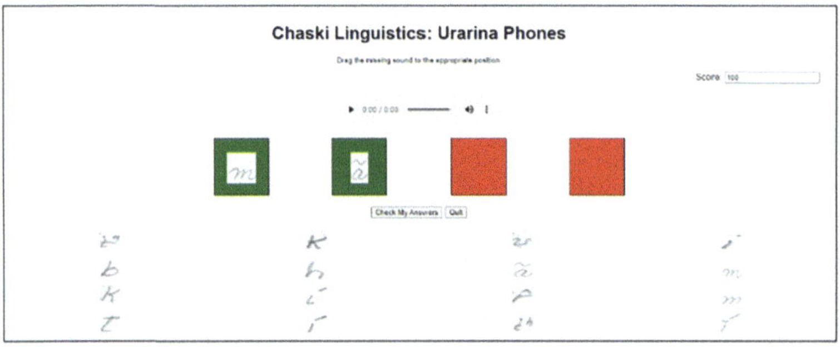

Fig. 12. Linguistics Phone Game User Interface

3 Conclusion

This project has shown the process of restoring and preserving handwritten notes from a linguistics field study that occurred during the 1980s, using machine learning. As part of this project, more than 2,000 lexicon cards were digitized. Employing Google Cloud Vision and manual efforts, the phonetic spelling of words was extracted and then further segmented into character images. Google Cloud Vision was also utilized to identify descriptive text written in Spanish. With the extracted characters, a CNN-based application was created to identify and classify the characters of the lexicon cards. Once all the Spanish text and phonetic characters were recognized, an editing system allowed a human to revise the results and create a data store with the final human-confirmed data. For the final data verification, a purpose-built game challenged players to identify phones within a recording. Although gaming in academic environments cannot be considered a new idea [4,11], using games to educate and review data in a linguistics environment is a novel concept. Games designed in this project provide valuable data gathered to determine if a card needs expert review, and gaming enables players to practice phone identification. This project has established a framework for the restoration of archival materials from a linguistics study and is envisioned to aid researchers in revitalizing comparable projects and preserving additional endangered languages.

Acknowledgements. The authors wish to thank Anders Koskinen for his incredible efforts in reviewing this document.

References

1. Cajas Rojas, J., Corbera, A., Gualdieri, B., Solis, G.: Bibliografía etnolingüística urarina. Instituto de Lingüística Aplicada (CILA), Universidad Nacional Mayor de San Marcos, Lima (1987)
2. Cameron, P., Schoen, R.: Building an automation center of excellence. Fortra Blog (2017). https://www.fortra.com/blog/building-automation-center-excellence-part-two-scalable-rpa-infrastructure
3. Cloud, S.: How to load PyTorch Dataloader into GPU — Saturn Cloud Blog — saturncloud.io. https://saturncloud.io/blog/how-to-load-pytorch-dataloader-into-gpu/. Accessed 09 Sep 2024
4. Cohen, K.J., Rhenman, E.: The role of management games in education and research. Manag. Sci. **7**(2), 131–166 (1961). https://www.jstor.org/stable/2627098
5. Datta, K., et al.: Training multiscale-CNN for large microscopy image classification in one hour. In: Weiland, M., Juckeland, G., Alam, S., Jagode, H. (eds.) ISC High Performance 2019. LNCS, vol. 11887, pp. 463–477. Springer, Cham (2019). https://doi.org/10.1007/978-3-030-34356-9_35
6. Dhillon, C.W.A.: Character recognition using python — handwritten character recognition using OpenCV and sklearn (2021). https://youtu.be/Sc_ycuEoGQw
7. Dumoulin, V., Visin, F.: A guide to convolution arithmetic for deep learning. arXiv preprint arXiv:1603.07285 (2016)
8. El-Haj, M., Kruschwitz, U., Fox, C.: Creating language resources for under-resourced languages: methodologies, and experiments with Arabic. Lang. Resour. Eval. **49**(3), 549–580 (2014). https://doi.org/10.1007/s10579-014-9274-3
9. Fujitsu: Fujitsu fi-7160 scanner. https://www.pfu.ricoh.com/%20global/products/computing/peripheral/scanners/fi7160/. Accessed 20 July 2024
10. Glorot, X., Bordes, A., Bengio, Y.: Deep sparse rectifier neural networks. In: Proceedings of the Fourteenth International Conference on Artificial Intelligence and Statistics, pp. 315–323 (2011). https://proceedings.mlr.press/v15/glorot11a/glorot11a.pdf
11. Greenblat, C.S.: Teaching with simulation games: a review of claims and evidence. Teach. Sociol. **1**(1), 62–83 (1973). https://doi.org/10.2307/1317334

12. Heydegger, V.: Analysing the impact of file formats on data integrity. In: Archiving Conference, vol. 2008, pp. 50–55. Society for Imaging Science and Technology (2008). https://image.dig4e.com/IS&T%20Articles/arch2008-s11LDC-Heydegger.pdf
13. Ioffe, S., Szegedy, C.: Batch normalization: accelerating deep network training by reducing internal covariate shift. In: International Conference on Machine Learning, pp. 448–456. PMLR (2015)
14. Jin, Y., et al.: Efficient multimodal large language models: a survey. arXiv preprint arXiv:2405.10739 (2024)
15. Keras: Flatten layer (2023). https://keras.io/api/layers/reshaping_layers/flatten/
16. Krizhevsky, A., Sutskever, I., Hinton, G.E.: ImageNet classification with deep convolutional neural networks. Commun. ACM **60**(6), 84–90 (2017). https://doi.org/10.1145/3065386
17. Krystalli, P.: Using educational online games to develop linguistic competences in foreign language learning. In: INTED2015 Proceedings, pp. 5338–5346. IATED (2015)
18. Ladefoged, P., Johnson, K.: A Course in Phonetics. Thomson Wadsworth (2006)
19. Mamun, F., Chowdhury, S., Giti, J., Sarker, H.: Classification of non-native handwritten characters using convolutional neural network. arXiv preprint arXiv:2406.04511 (2024)
20. Mao, A., Mohri, M., Zhong, Y.: Cross-entropy loss functions: theoretical analysis and applications. In: International Conference on Machine Learning, pp. 23803–23828. PMLR (2023). https://proceedings.mlr.press/v202/mao23b.html
21. Matyas, S., Matyas, C., Schlieder, C., Kiefer, P., Mitarai, H., Kamata, M.: Designing location-based mobile games with a purpose - collecting geospatial data with cityexplorer. In: Proceedings of the 2008 International Conference on Advances in Computer Entertainment Technology, ACE 2008, pp. 244–247 (2008). https://doi.org/10.1145/1501750.1501806
22. McConvell, P., Day, R., Black, P., et al.: Making a Meriam Mir dictionary. In: Papers in Australian Linguistics No. 15: Australian Aboriginal lexicography. Pacific Linguistics (1983)
23. Microsoft: Azure cognitive services computer vision. https://azure.microsoft.com/en-us/services/cognitive-services/computer-vision/
24. Mishra, M.: Convolutional neural networks, explained. Towards Data Science (2020). https://towardsdatascience.com/convolutional-neural-networks-explained-9cc5188c4939
25. Narayan, A., Muthalagu, R.: Image character recognition using convolutional neural networks. In: 2021 Seventh International conference on Bio Signals, Images, and Instrumentation (ICBSII), pp. 1–5. IEEE (2021). https://doi.org/10.1109/ICBSII51839.2021.9445136
26. Pasero, R., Sabatier, P.: Linguistic games for language learning: a special use of the illico library. Comput. Assist. Lang. Learn. **11**(5), 561–585 (1998). https://doi.org/10.1076/call.11.5.561.5663
27. Platform, G.C.: Cloud vision API. https://cloud.google.com/vision/docs/ocr
28. Postindustria: How much data is required for machine learning (2024). https://postindustria.com/how-much-data-is-required-for-machine-learning/. Accessed 20 July 2024
29. Saarenpää, H.: Data gathering methods for evaluating playability of pervasive mobile games (2008)

30. Sarika, N., Sirisala, N., Velpuru, M.S.: CNN based optical character recognition and applications. In: 2021 6th International Conference on Inventive Computation Technologies (ICICT), pp. 666–672. IEEE (2021). https://doi.org/10.1109/icict50816.2021.9358735
31. Google Scholar: A Google scholar search for linguistics research between 1900 and 1960 conducted on Google scholar. https://scholar.google.com/
32. Amazon Web Services: Amazon textract. https://aws.amazon.com/textract/
33. Shorten, C., Khoshgoftaar, T.M.: A survey on image data augmentation for deep learning. J. Big Data **6**(1), 1–48 (2019). https://doi.org/10.1186/s40537-019-0197-0
34. Smith, L.N.: Cyclical learning rates for training neural networks. In: 2017 IEEE Winter Conference on Applications of Computer Vision (WACV), pp. 464–472. IEEE (2017). https://doi.org/10.1109/WACV.2017.58
35. Smith, R.: Tesseract OCR (2006). https://github.com/tesseract-ocr/tesseract
36. Sunubi, A.H.: Introduction to Linguistics. DIRAH, Parepare, Indonesia (2016)
37. UNESCO: Urarina. https://en.wal.unesco.org/languages/urarina. Accessed 29 June 2024
38. WWT: How to automate everything right: WWT's 6 principles of automation. WWT Website (2024). https://www.wwt.com/article/6-principles-of-automation
39. Yu, W., Liu, Y., Hua, W., Jiang, D., Ren, B., Bai, X.: Turning a clip model into a scene text detector. In: Proceedings of the IEEE/CVF Conference on Computer Vision and Pattern Recognition, pp. 6978–6988 (2023)
40. Zumarnis, M.R., Purwati, P.D., et al.: The effectiveness of lexicon card media assisted by minimum competency assessment on students' learning comprehension. J. Ilmiah Sekolah Dasar **8**(2) (2024). https://doi.org/10.23887/jisd.v8i2.70587

Implementing Adaptability Techniques on a Serious Game for Children and Adolescents with ASD

Guilherme Nunes Azevedo[1(✉)], Mariane Gomes Lacerda[2], Luciana Correia Lima De Faria Borges[1], and Eunice Pereira Dos Santos Nunes[1]

[1] Universidade Federal de Mato Grosso, Instituto de Computação, Cuiabá, Mato Grosso, Brazil
guilherme.azevedo@sou.ufmt.br, lucianafariaborges@gmail.com, eunice.nunes@ufmt.br

[2] Universidade Federal de Mato Grosso, Mestrado Profissional em Educação Inclusiva em Rede Nacional, Cuiabá, Mato Grosso, Brazil
mariane.gomes@edu.se.df.gov.br

Abstract. This study investigates the implementation of new adaptive tools in a serious game for children and adolescents with autism spectrum disorder (ASD). The aim is to evaluate how these adaptations, initially made with physical materials and later implemented in the digital game, can improve the players' experience and performance. The research was justified by the need to customize games to meet the specific needs of children and adolescents with ASD. The methodology included the application of neuropsychological tests and the use of tablets, PCs and notebooks for interventions. The results indicate significant improvements in the participants' cognitive and attention skills. We conclude that adaptability can be an effective tool in the development of games for this audience.

Keywords: serious game · ASD · adaptability · personalized learning

1 Introduction

Children and adolescents with autism spectrum disorder (ASD) face unique challenges in social interaction, communication, and information processing. These difficulties can limit their participation in playful activities, restricting their experiences and developmental opportunities [20].

However, studies indicate that digital games play a significant role in improving cognitive and social skills, as well as promoting inclusion and interaction with others [7,10,16]. These findings highlight the potential of digital games as effective tools in supporting the development of children and adolescents with ASD.

F. Suni-Lopez et al. (Eds.): CIIS 2024, ISEM 48, pp. 18–30, 2025.
https://doi.org/10.1007/978-3-031-91481-2_2

Digital games, defined as electronic games that leverage graphical interfaces, stand out for their ability to create interactive and engaging environments. In addition to being widely recognized in the entertainment landscape, they have been relevantly integrated into education and therapy. Unlike traditional games, which rely on physical elements like boards and cards, digital games use electronic devices such as computers, consoles, smartphones, and tablets, providing a more immersive and dynamic experience [21].

Studies, such as those by [8,14], highlight the possibility of digital games to be customized to meet specific demands. One example is the Hope Project, which uses simulations to teach the Turkish language and other skills to Syrian refugee children, demonstrating the educational and therapeutic applicability of simulation [19]. These studies indicate that digital games have the potential not only to teach academic content, but also to enhance cognitive development, especially in areas like attention and efficiency, through a playful approach.

Although digital games are a tool to help treat children and adolescents with ASD, individuals with autism have unique sets of challenges that must be addressed to contribute to their development and integration into society [12]. To do this, games need to be adaptable to meet the specific needs of each patient.

To assist this audience, the *Laboratório de Ambientes Virtuais Interativos (LAVI)* research group, affiliated with the *Universidade Federal de Mato Grosso*, is developing the game *Dividir e Conquistar*. The group is made up of a teacher who specializes in virtual reality, a teacher who specializes in Human-Computer Interaction, a teacher and researcher in the field of ASD, and an undergraduate student of information systems. This is a serious game designed to teach or train skills using motivational principles from conventional games for a serious purpose [13].

The game adopts a specific adaptability approach to meet the individual needs of autistic players, aiming to provide a more inclusive and beneficial experience by considering aspects such as social interaction, communication, and information processing [4,7,11,22]. This game aims to stimulate collaboration and train fine motor coordination in children and adolescents with ASD, using adaptive difficulty applied to a puzzle.

To develop this game, software designed for game creation called Unity was used together with the C# programming language. The student had a work plan defined together with the teachers, and, in this plan there were deadlines and functionalities to be implemented in the game. From this, meetings were held so that the student could show the progress of their work and the features developed in the game. During these meetings, validations and suggestions were made by teachers regarding the game to assist and guide the development process.

The game involves solving puzzles in pairs before time runs out. It is in the final stage of development and is based on two systematic reviews made by the research group mentioned earlier.

The aim is therefore to investigate the effectiveness of new adaptive tools implemented in the serious game *Dividir e Conquistar* for children and adolescents with ASD.

2 Methodology

This study was designed as exploratory research, approved by the Human Research Ethics Committee under CAAE number 71225823.0.0000.5690, with the primary objective of evaluating the effectiveness of adaptations implemented in a serious game developed for children and adolescents with ASD. The sample consisted of six children and adolescents with confirmed ASD diagnoses, including three participants aged 11 and three aged 12, with 67% being male and 33% female. Participants were selected based on the confirmation of their ASD diagnosis and their ability to participate in the game sessions. Although the number of participants is limited, the study aims to provide initial insights, recognizing the challenges inherent in conducting interventions with this population.

The interventions took place during the first week of November 2023, consisting of two 30-minute sessions each. Before starting the game activities, participants underwent two neuropsychological tests: the Tower of London test (ToL) and the attention cancellation test (TAC), chosen to assess planning, problem-solving, and selective attention skills. The tests were administered before and after the interventions, allowing for a comparison of results.

The sessions began with a brief discussion about the players' previous experiences with puzzles or similar games. The second author, the participants' specialized educational assistance teacher, presented the game's objectives and contextualized the activity. During the game exploration, the researcher demonstrated critical thinking and problem-solving strategies, providing specific feedback and individualized guidance to encourage players to reflect on their decisions. After completing the game, a reflective discussion took place, during which players shared their strategies and discussed how they could apply them in other situations.

For the interventions, Samsung Galaxy Tab tablets, a PC, and a notebook were used, ensuring consistency in game conditions. During the sessions, the need for physical materials, such as printed images of the puzzles, was observed to help the players when necessary.

Data were collected through direct observations conducted by the second author. The information obtained was organized into Excel spreadsheets and analyzed using Jamovi software, version 2.4. Various statistical techniques were employed, including the Student's t-test for paired samples, the Wilcoxon test, and the Shapiro-Wilk normality test. Additionally, contingency tables and graphs were constructed to aid in data visualization. The analysis focused on identifying changes in participants' cognitive and attention skills, correlating these changes with the interventions performed.

3 Results and Discussions

3.1 Presentation of the Game *Dividir e Conquistar*

The game is designed for children and adolescents between the ages of 6 and 12 and aims to stimulate collaboration, communication, fine motor coordination training, and the enhancement of executive functions. It features a 3D virtual environment (VE) representing a school and a collaborative puzzle minigame. The game offers an engaging narrative in a minimalist setting, with multiplayer features to promote social interaction. Studies indicate that experiences in 3D VEs can improve the expressiveness of autistic children [17,25]).

At the beginning of the game, the player chooses an avatar from options that include different ethnicities and persons with disabilities (PwD). The game starts in a school context, where the player interacts with the teacher to start the minigame as shown in Fig. 1. The objective is to complete the puzzle within the time limit by discussing strategies in a turn-based system. Puzzle images are randomly selected to keep the game dynamic. If the pair completes the puzzle within the time, they are congratulated; otherwise, they are encouraged to try again.

Fig. 1. Player talking with the teacher in the 3D virtual environment

Auditory, visual, and haptic feedback are provided when the player interacts with the puzzle pieces. A help button appears after a period of no progress, assisting in puzzle assembly.

Despite all these features already incorporated into the game, there is a need for more efficient techniques focused on game adaptability. Combined with the collaborative aspect already in use, it is believed that this could greatly assist in overcoming social barriers, developing essential skills in children and adolescents with ASD.

To develop the game, several systematic reviews were made by the research group to understand the needs of the target audience. From this, the game's resources were developed based on the data returned from these reviews so that the game could be an effective tool during therapy sessions for children and adolescents with ASD.

3.2 Adaptability in Games

In this context, adaptability refers to features in serious games that can be changed either by the game itself or by the user [2]. This approach allows the game to be more inclusive for players and allows the therapist to have control over the intervention and be able to change it according to the needs of each patient and situation.

Digital games often offer practical mechanisms such as real-time feedback and difficulty-level adjustment based on player performance. This approach optimizes the learning process, as evidenced by recent research [5,6,9,15,24]. Real-time feedback, combined with adaptability, creates a dynamic learning experience allowing the player to progress at their own pace and gradually overcome challenges.

Research by [22] has demonstrated the effectiveness of personalized games in improving working memory and reducing autistic symptoms in children. [10] compared virtual training with physical exercises, revealing that virtual training with adapted digital games achieved more positive results. [16] used digital games in therapeutic sessions to improve executive functions such as attention, working memory, and cognitive flexibility, achieving significant improvements.

The study implemented various adaptability techniques in the game *Dividir e Conquistar*, initially using physical materials, such as printed out images of the puzzles, to visually assist players who needed help. These observations were provided by the second author, a specialized care teacher, who saw the need to adapt the material to better meet the needs of players with ASD. The recommendations made through meetings with the group of developers resulted in the insertion of new adaptability features into the digital game.

In this way, the implementation of the adaptability features was based on gaps identified during testing and suggestions made by members of the research group. In addition, adaptability techniques were also identified through a literature review and implemented. While the review was being carried out, a test was performed to assess the effectiveness of the game as a tool for encouraging the collaboration of children with ASD, taking into account the personalization aspects for the target audience.

3.3 Adaptability Features

Firstly, a reference image was added for the construction of the puzzle; this image can be activated or deactivated based on the player's choice. The image is on the left side of the device, which allows the player to know what the complete image looks like during puzzle solving, as shown in Fig. 2. The aim of this functionality is to visually assist players who needed help understanding what the final image would look like after solving the puzzle.

Another feature implemented was to allow the user to add extra images to the game. From this, a feature was implemented that receives a link to an image hosted on the web, imports the image as a texture into the game, and separates the imported image into pieces. Therefore, it is possible to implement new images

Fig. 2. Reference image feature

for the game from a link on the web, as illustrated in Figs. 3 and 4. The purpose with this functionality is to increase players' level of engagement by allowing them to insert any image they wish to solve in the puzzle.

Fig. 3. Extra image feature (screen for placing the image link)

3.4 Implementation of Adaptability Techniques

Personalization of the media used in the game [1,3,23] and allowing the game's difficulty to be adapted by the player [3,12] were identified as adaptability techniques. Based on this, these techniques were implemented with the following

Fig. 4. Extra image feature (screen after loading the image)

functionalities: buttons that allow the player to change the visual and auditory feedback received when correctly fitting a piece, and buttons that allow the player to select the number of pieces for the puzzle (see Fig. 5).

While solving the puzzle, the player can access a screen where they can change a series of options for executing the puzzle. Among these options, he will be able to change the image that appears as feedback when he places the piece in the correct location on the board and the audio that is played when this happens, as shown in Fig. 6.

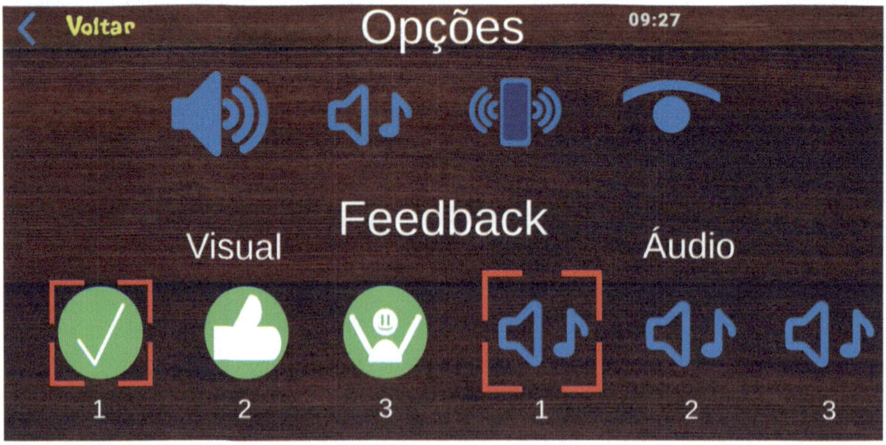

Fig. 5. Feedback customization

When players gather in the virtual environment to play the cooperative puzzle, one of them has the possibility to choose which image will be assembled, as shown in Fig. 2. After that, it is possible to define the number of pieces that the puzzle to be solved will have, as shown in Fig. 7. It is possible to define the puzzle as having 9, 16 or 36 pieces.

Fig. 6. Puzzle selection screen

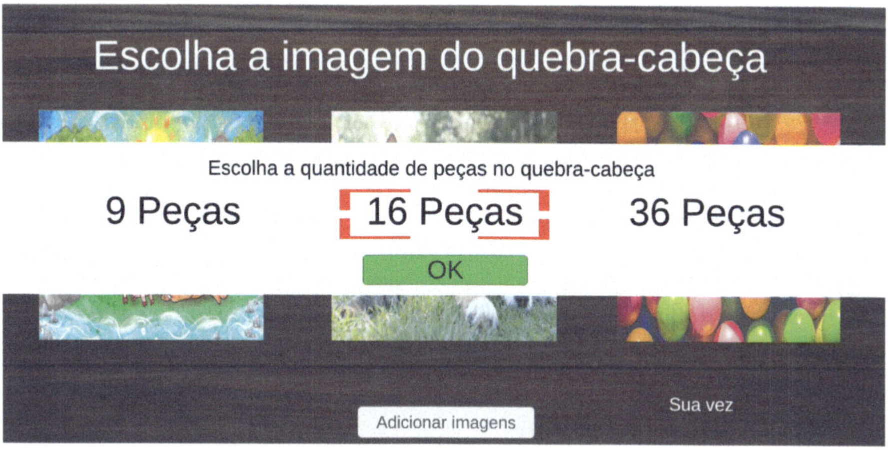

Fig. 7. Puzzle piece quantity selection screen

3.5 Collected Data

A thorough analysis of the collected data is crucial to understand the impact of digital game interventions on the cognitive development of children and adolescents with ASD. This study evaluated six participants, all between 11 and 12 years old, using two main neuropsychological tests: ToL and TAC.

As described in the book *Avaliação Neuropsicológica Cognitiva - Atenção e Funções Executivas*, organized by [18], the ToL, developed by Shallice in 1982, is a widely used tool for assessing executive functions, specifically the ability to plan and solve problems. The test consists of a task in which the participant must move balls of different colors between three pegs to reach a target configuration, following specific rules. Performance in the ToL is measured by the number of moves made, the time taken to complete the task, and the number of errors committed.

In the present study, the descriptive results before the intervention showed that the average scores of the participants were 91.7 (standard deviation of 19.9), indicating significant variability in initial performances. After the interventions with the game *Dividir e Conquistar*, the average scores increased to 103.7. Despite this apparent improvement, statistical tests such as Student's t-test ($p = 0.153$) and the Wilcoxon test ($p = 0.281$) did not reveal statistically significant differences. The normality analysis of the data using the Shapiro-Wilk test ($p = 0.911$) suggests that the data follows a normal distribution. However, the lack of statistical significance may be attributed to the limited sample size ($n = 6$), the short intervention period, and the diversity of skills among participants, highlighting the need for a larger sample and a longer intervention period to validate these findings.

The TAC, also detailed by [18], assesses sustained and selective attention. During the test, participants must locate and mark specific target stimuli on a sheet filled with distractors, with results measured by the total completion time and accuracy, including omission and commission errors.

Before the intervention, the average scores in the TAC were 74.5, with a median of 76.0. After the intervention, there was a significant increase in the average to 87.3 and the median to 85.0. This change was confirmed to be statistically significant by the paired sample t-test ($t = -3.59$, $p = 0.016$), demonstrating a substantial improvement in the selective attention of the participants. The effect size analysis (Cohen's $d = -1.47$) reinforces the robustness of this impact. The distribution of scores after the intervention remained approximately normal, according to the Shapiro-Wilk test analysis, which strengthens the validity of the obtained results.

3.6 Qualitative Observations

In addition to the quantitative data, qualitative observations made during the intervention sessions provided valuable information about the impact of the interventions on the participants:

- Player 1, aged 11, stood out for exceptional collaboration and performance, suggesting that the use of the game *Dividir e Conquistar* can be an effective tool for developing problem-solving and planning skills in adolescents with ASD.

- Player 2, aged 12, showed significant improvements in cognitive and motivational skills, including strategic management and problem-solving competencies.
- Player 3, aged 11, faced persistent engagement and motivation challenges, indicating the need for more individualized intervention approaches to enhance their engagement.
- Player 4, also aged 11, showed a consistent trajectory of success, with superior cognitive abilities and high motivation.
- Player 5 continued to face significant challenges, highlighting the need for more specific motivational strategies.
- Player 6, aged 12, demonstrated consistent progress and high performance, particularly in collaborative strategies.

These qualitative observations emphasize the importance of tailoring intervention strategies to the specific needs and characteristics of each participant, highlighting the diversity of responses within the population of adolescents with ASD.

3.7 Methodological Considerations and Prospects for Expansion

This study presents promising results; however, an important limitation lies in the small sample size, consisting of six participants, which compromises the generalizability of the findings. To strengthen external validity, it is recommended to increase the number of participants and include a broader age range, from 6 to 12 years old, as outlined in the initial planning. Such an adjustment would allow for a more representative analysis of the effects of interventions with the game *Dividir e Conquistar*.

As well as to increasing the sample size, another relevant aspect for methodological improvement is the incorporation of additional metrics and the use of more robust statistical approaches, which allow for a more in-depth analysis of the quantitative data, ensuring greater analytical rigor.

Finally, the adoption of a case study design would allow for exploring the individual experiences of the participants in more detail, enriching the qualitative analysis and offering a deeper understanding of the particularities of the behavioral and cognitive responses to the game. This approach would make it possible to complement quantitative data, avoiding undue generalizations and providing more detailed information on the impact of interventions.

These expansion and improvement strategies would address the identified limitations, contributing to the advancement of knowledge on using the game *Dividir e Conquistar* as an educational tool for children and adolescents with ASD.

4 Conclusions

Digital games represent a valuable resource, offering an engaging approach to the development of children and adolescents with ASD. The ability to adapt, personalize, and provide real-time feedback makes these games powerful tools for enhancing social and cognitive skills. However, it is essential that these interventions are personalized and continuously evaluated to maximize their benefits.

Therefore, a test was carried out to evaluate the effectiveness of the *Dividir e Conquistar* game in encouraging collaboration among children with ASD, taking into account the aspects of personalization for the target audience.

Furthermore, in this paper the following adaptability techniques for game based 3D VEs were identified: personalization of the media used in the game, and allowing the game's difficulty to be adapted by the player. These techniques were implemented in the form of buttons that allow the player to change the visual and auditory feedback received when correctly fitting a piece and buttons that allow the player to select the number of pieces for the puzzle.

Moreover, the following adaptability features were presented: reference images to support puzzle solving, and allowing the user to add extra images to the game. Therefore, it is expected that after identifying the previously mentioned adaptability techniques and implementing them, the game can be a tool for therapy sessions, promoting the inclusion of children and adolescents with ASD.

The game *Dividir e Conquistar* stands out as a promising tool, combining collaborative and adaptability elements to meet the individual needs of this audience. Future studies should further explore the effectiveness of these approaches, ensuring that each intervention is adjusted according to the unique characteristics of each player.

Acknowledgments. The authors thank the *Fundação de Amparo à Pesquisa do Estado de Mato Grosso (FAPEMAT)*, and the Pro-Rectorate of Research of *Universidade Federal de Mato Grosso*.

References

1. Aguiar, Y., Galy, E., Godde, A., Trémaud, M., Tardif, C.: AutismGuide: a usability guidelines to design software solutions for users with autism spectrum disorder. Behav. Inf. Technol. **41**(6), 1132–1150 (2022). https://doi.org/10.1080/0144929X.2020.1856927
2. Aranha, R.V.: Adaptação automática de jogos sérios a partir de emoções e traços de personalidade. Tese (Doutorado) - Escola Politécnica da Universidade de São Paulo. Departamento de Engenharia de Computação e Sistemas Digitais (2022)
3. Carvalho, A., Braz, C.S., Santos, S., Ferreira, R., Prates, R.O.: Serious games for children with autism spectrum disorder: a systematic literature review. Int. J. Hum.-Comput. Interact. (2023). https://doi.org/10.1080/10447318.2023.2194051
4. Carvalho, C.P., Santos, S.S., Escudeiro, G.M., Pinheiro, K.V., Neto, N.C.S., Mota, M.P.: A study on customizing interaction in adaptable games. In: Proceedings of the 18th Brazilian Symposium on Human Factors in Computing Systems (IHC '19). Association for Computing Machinery, New York, NY, USA, Article 6, pp. 1–11 (2019). https://doi.org/10.1145/3357155.3358468

5. Faja, S., et al.: A preliminary randomized, controlled trial of executive function training for children with autism spectrum disorder. Autism **26**(2), 346–360 (2022)
6. Fernandes, M., Nohama, P.: Jogos Digitais para Pessoas com Transtornos do Espectro do Autismo (TEA): Uma Revisão Sistemática. Revi. Iberoamericana Tecnología en Educación y Educación en Tecnología (26), 72–80 (2020)
7. Ferreira, E., Oliveira, C., Moreira, O., Pereira, E.: Effect of exergames on motor performance and reaction time in a child with autism spectrum disorder / Efeito do exergames no desempenho motor e no tempo de reação em uma criança com transtorno do espectro autista. Motricidade, 1–14 (2022). https://doi.org/10.6063/motricidade.27119
8. Homer, B.D., et al.: Improving high school students' executive functions through digital game play. Comput. Educ. **117**, 50–58 (2018)
9. Hoseini, F.S., Khodadadi, M., Khorambakht, A.: The effect of 2D and 3D action video game interventions on executive functions in male students. Simul. Gaming **53**(5), 405–422 (2022)
10. Ji, C., Yang, J., Lin, L., Chen, S.: Executive function improvement for children with autism spectrum disorder: a comparative study between virtual training and physical exercise methods. Children **9**, 4 (2022). https://doi.org/10.3390/children9040507
11. Khan, A.M., Arsov, I., Preda, M., Chabridon, S., Beugnard, A.: Adaptable client-server architecture for mobile multiplayer games. In: Proceedings of the 3rd International ICST Conference on Simulation Tools and Techniques (SIMUTools '10). ICST (Institute for Computer Sciences, Social-Informatics and Telecommunications Engineering), Brussels, BEL, Article 11, pp. 1–7 (2010). https://doi.org/10.4108/ICST.SIMUTOOLS2010.8704
12. Lopez-Herrejon, R.E., Poddar, O., Herrera, G., Sevilla, J.: Customization support in computer-based technologies for autism: a systematic mapping study. Int. J. Hum.-Comput. Interact. **36**(13), 1273–1290 (2020). https://doi.org/10.1080/10447318.2020.1731673
13. Mert, A., Nijboer, T., Doyen, B., Meijer, H., Dankbaar, M.: A medical serious games framework hierarchy for validity. In: 10th International Conference on Virtual Worlds and Games for Serious Applications (VS-Games), Würzburg, Germany, pp. 1–2 (2018). https://doi.org/10.1109/VS-Games.2018.8493405
14. Plass, J.L., et al.: The effect of adaptive difficulty adjustment on the effectiveness of a game to develop executive function skills for learners of different ages. Cogn. Dev. **49**, 56–67 (2019)
15. Torske, T., et al.: Autism spectrum disorder polygenic scores are associated with every day executive function in children admitted for clinical assessment. Autism Res. **13**(2), 207–220 (2020)
16. Ramos, D.K., Venturiei, C., Stange, N., Anastácio, B.S.: Jogos digitais e gamificação: intervenção para o aprimoramento das funções executivas. Debates em Educação **12**, 499–520 (2020). https://doi.org/10.28998/2175-6600.2020v12n27p499-520
17. Ringland, K. E.: Do you work for aperture science? Researching and finding the gamer identity in a Minecraft community for autistic children. In: FDG'19, San Luis Obispo, CA, USA (2019)
18. Seabra, A.G., Dias, N.M.: Avaliação neuropsicológica cognitiva: atenção e funções executivas. São Paulo: Mennon Edições Científicas **1**, 174 (2012)
19. Sirin, S., et al.: Digital game-based education for Syrian refugee children: project hope. Vulnerable Child. Youth Stud. **13**(1), 7–18 (2018)

20. Sulkes, S.B.: Autism Spectrum Disorder (2022). https://www.msdmanuals.com/professional/pediatrics/learning-and-developmental-disorders/autism-spectrum-disorders
21. Vajawat, B., Varshney, P., Banerjee, D.: Digital gaming interventions in psychiatry: evidence, applications and challenges. Psychiatry Res. **295**, 113585 (2021)
22. Wagle, S., et al.: Development and testing of a game-based digital intervention for working memory training in autism spectrum disorder. Sci. Rep. **11**(1), 13800 (2021). https://doi.org/10.1038/s41598-021-93258-w
23. Weisblatt, E.J., et al.: A tablet computer-assisted motor and language skills training program to promote communication development in children with autism: development and pilot study. Int. J. Hum.-Comput. Interact. **35**(8), 643–665 (2019). https://doi.org/10.1080/10447318.2018.1550176
24. Weismer, S.E., et al.: Executive function skills in school-age children with autism spectrum disorder: association with language abilities. J. Speech Lang. Hear. Res. **61**(11), 2641–2658 (2018)
25. Wu, Q., et al.: Squeeze the ball: designing an interactive playground towards aiding social activities of children with low-function autism. In: CHI 2020, Honolulu, HI, USA (2020)

CuidaMED: An Artificially Intelligent and Serious Game Platform for Chronic Patient's Adherence Powered by Gamification

José Vallés Bedregal and Jean Pierre Vallés Bedregal(✉)

Universidad de Lima, Avenida Javier Prado Este 4600, Santiago de Surco, Lima, Peru
jvallesb@alumni.nd.edu

Abstract. Games are increasingly recognized as a potent tool for motivating behavioral change, particularly in healthcare. This study introduces CuidaMED, a novel piece of software designed to assist patients with chronic conditions in adhering to their treatments and managing their diseases. CuidaMED leverages serious games-interactive platforms with goals beyond mere entertainment-incorporating gamification strategies and a robust technical architecture to foster positive behavioral changes. Beyond its gamified design, CuidaMED stands out by incorporating the latest AI interaction patterns, specifically large language models, to understand and respond to patient needs through natural human language. This allows for the dynamic generation of personalized games that adapt to medium-term healthcare needs, addressing the limitations of static health apps-which often feature static content-with dynamic, adaptive content. Powered by generative AI, these games maintain patient engagement with educational content and treatment adherence. The platform also features a virtual companion, Mary, who provides contextual rewards to encourage patients to follow their treatment plans, reinforcing positive behaviors. Additionally, CuidaMED provides real-time reports to family members and caregivers, alerting them to any significant changes in the patient's condition. By leveraging the latest AI interaction patterns, including large language models and generative AI, CuidaMED dynamically adapts games to patient needs, enhancing engagement and personalizing treatment adherence. This study explores the design, implementation, and efficacy of CuidaMED, presenting initial results that demonstrate its potential to improve patient engagement and treatment adherence through innovative, AI-powered solutions. The findings highlight the significance of gamification, adaptive content, and serious games in healthcare, supporting patients in achieving better health outcomes over time.

Keywords: serious games · healthcare · chronic conditions · behavior change · gamification · patient adherence · generative AIai · large language models (LLMs)

F. Suni-Lopez et al. (Eds.): CIIS 2024, ISEM 48, pp. 31–50, 2025.
https://doi.org/10.1007/978-3-031-91481-2_3

1 Introduction

In the global health landscape, patient non-adherence-particularly among individuals with chronic illnesses-has become a critical challenge, significantly impacting health outcomes and escalating healthcare costs worldwide. Nearly 50% of chronic patients discontinue their treatments within the first year, resulting in alarming consequences, including hundreds of thousands of deaths annually and healthcare expenditures amounting to billions of dollars [4,18,20]. This issue is especially acute in Peru, where health service institutions (IPRESS) report an average consultation time of just 12 min, with patients enduring waiting periods ranging from 35 to 114 min across different facilities [13,22]. Addressing these challenges requires innovative solutions that harness modern technology and insights from behavioral science.

CuidaMED represents a pioneering approach to enhancing treatment adherence and disease management by integrating artificial intelligence, natural human language processing, gamification, and serious games. The platform transforms the typically challenging task of chronic disease management into an engaging and rewarding experience through its dynamic and adaptive games. CuidaMED features a virtual companion, Mary, who provides personalized reminders, notifications, and connects users with healthcare professionals when necessary. Utilizing principles from behavioral economics and gamification, such as the Fogg behavior model [12] and the Octalysis framework [8], CuidaMED aims to sustain patient motivation and facilitate the formation of healthy habits. By addressing the gaps in existing health applications-namely, their reliance on static content- CuidaMED offers a dynamic, AI-driven approach to patient engagement and treatment adherence by leveraging natural human language processing to personalize interactions and tailor the patient experience. This paper presents an in-depth examination of CuidaMED's design, implementation, and preliminary efficacy. It details the platform's technological architecture, its development process, and findings from initial usability testing and pilot studies. The goal is to demonstrate CuidaMED's potential to significantly improve patient adherence and health outcomes by integrating behavioral design and cutting-edge technology into healthcare applications.

This paper is structured into six sections. Section 2, Habits and Behavioral Economics, lays the groundwork for understanding the behavioral patterns that influence patient adherence. Section 3, Gamification and Behavioral Design, examines the principles underpinning CuidaMED's strategies for user engagement. Section 4, Software Architecture and Interoperability, describes the technological framework that enables the platform's adaptive functionalities. Section 5, Effectiveness of CuidaMED, presents findings from initial usability tests and pilot studies to evaluate the platform's impact. Section 6, Security and Privacy, focuses on the measures implemented to ensure the protection of patient data. Finally, the paper concludes by summarizing key findings and outlining future directions.

2 Habits and Behavioral Economics

In the global health context, patient non-adherence to medical treatments, particularly among those with chronic illnesses, poses a significant challenge with severe consequences. To devise effective solutions, we must first understand why individuals often fail to act in their own best interests. Behavioral economics, merging insights from economics and psychology, provides a framework for understanding these seemingly irrational decisions. Unlike neoclassical economics, which assumes rational, well-informed decision-making, behavioral economics reveals that people are predictably irrational [3]. By leveraging principles such as loss aversion, present bias, and the intent-behavior gap, we can design interventions that align with actual human behavior.

The intent-behavior gap illustrates the discrepancy between intentions and actions-patients may plan to follow their treatment plans but often fail to do so due to immediate distractions or competing priorities. Present bias highlights the human tendency to prioritize immediate gratification over long-term benefits, leading patients to skip medications that do not provide instant relief. Loss aversion suggests that patients are more motivated by the fear of losing something they already have than by potential gains, making upfront incentives effective in encouraging adherence.

Behavioral design theory offers practical strategies for changing habits, which is essential for improving patient adherence to treatment plans. The Fogg behavior model, for instance, highlights that behavior change relies on the interplay of motivation, ability, and triggers. The model shows that simplifying tasks increases the likelihood of adherence, even with lower motivation [8].

The habit loop theory further breaks down habit formation into four stages: cue, crave, response, and reward, emphasizing the importance of positive reinforcement in maintaining new behaviors [11].

According to behavioral economist Richard Thaler [23], the rewards themselves do not drive habit-forming actions in the long term. Long-term habit formation is achieved through doing what Thaler calls "small stuff" often enough to learn to get it right. This process requires practice. Thaler explains, "We do small stuff often enough to learn to get it right. Because learning takes practice, we are more likely to get things right at small stakes." [9]. Supporting this view, research from the University of Pennsylvania's Kevin Volpp, MD, demonstrates that effective incentive programs can offset present bias by providing rewards that make it more attractive to make healthy choices in the present [23].

Empirical evidence supports the effectiveness of these behavioral economics principles. One study demonstrated that providing incentives for 90 days significantly improved medication adherence, with benefits persisting three months after the incentives ceased [19]. Another study found that offering multiple incentives for gym attendance resulted in more durable exercise habits compared to a single incentive [21]. Additionally, a study focused on remote monitoring showed that incentives over 90 days led to lasting adherence to monitoring behaviors [6].

For over a decade, Volpp's Center for Health Incentives & Behavioral Economics has been studying the potential of behavioral economics to promote healthy actions like quitting smoking, losing weight, and taking medications. Their research has shown that programs with incentives-when applied in the right way and context-can help patients take an active role in their own healthcare.

These findings underscore the potential of behavioral economics and design to enhance health outcomes by fostering sustainable habits. By integrating these principles into solutions like CuidaMED, we can address non-adherence effectively, ultimately improving patient engagement and treatment success.

3 Gamification and Behavioral Design

3.1 Behavioral Design and Gamification

In non-game contexts, gamification is increasingly being adopted in the healthcare industry. This trend is motivated by the potential for gamification to engage patients in their own care, motivate healthy behavior, and improve health outcomes. Habit formation practices have been refined and made more effective through gamification, offering promising opportunities for chronic disease management.

As stated in the previous chapter, BJ Fogg's behavior model provides a foundation for understanding behavioral change [12]. The model is summarized by the equation:

$$B = MAT(P) \tag{1}$$

where behavior (B) is a function of motivation (M), ability (A), and triggers (T(P)). Each component must be present for behavior to occur. Motivation can be high or low, influenced by whether a task offers instant or delayed gratification. Ability refers to the simplicity or ease of performing a task. Triggers or prompts remind individuals to perform an action. Fogg identifies three types of triggers: signals, sparks, and facilitators. Signals are reminders for highly motivated individuals to perform a high-ability action. Sparks enhance motivation in people with high ability but low motivation. Facilitators aid those who are highly motivated to perform a low-ability action.

3.2 Gamification Design: Integrating Behavioral Design

In his book "Actionable Gamification: Beyond Points, Badges and Leaderboards," Chou (2019) [7] defines gamification as deriving the fun and addictive elements found in games and applying them to real-world activities. Chou describes the eight core drives of the Octalysis framework [8]:

Table 1 summarizes the eight core drives of the Octalysis framework, providing a clear and concise reference for integrating behavioral design in gamification.

Table 1. Summarizes the eight core drives of the Octalysis framework [8]

Core Drive	Description
Epic Meaning & Calling	The user feels privileged and unique, contributing to something larger.
Development & Accomplishment	The user's desire to excel and acquire new knowledge, driven by challenges.
Empowerment of Creativity & Feedback	The user feels autonomous, engaged in a creative process, and receives feedback.
Ownership & Possession	The user feels motivated by owning or controlling something, driving a desire to improve it.
Social Influence & Relatedness	Mentorship, collaboration, competition, social acceptance, and feedback motivate the user.
Scarcity & Impatience	The user values knowledge or actions by their rarity.
Unpredictability & Curiosity	The user's attention is kept through suspense and unexpected events.
Loss & Avoidance	The user is motivated to avoid negative outcomes, leveraging loss aversion.

3.3 Applying the Octalysis Framework

Using the Octalysis framework, a gamified application can be designed to leverage the Fogg behavior model for habit formation:

- **Motivation:** Core drives such as Epic Meaning & Calling, Development & Accomplishment, and Empowerment of Creativity & Feedback relate to gaining pleasure. Scarcity, Unpredictability, and Loss & Avoidance can create urgency to act, while Ownership & Possession and Social Influence & Relatedness enhance motivation.
- **Ability:** Simplicity can be influenced by core drives. For example, time constraints relate to Scarcity & Impatience and Loss & Avoidance, while social influences relate to Social Influence & Relatedness.
- **Triggers:** Triggers are crucial for habit formation and gamification. They can be signals, sparks, or facilitators, depending on the user's motivation and ability.

3.4 Gamification and Serious Games in Healthcare

An innovative extension of gamification is the concept of serious games, which are games designed for a primary purpose other than pure entertainment. Serious games have been successfully applied in healthcare to improve medication and treatment adherence. Research indicates that serious games can effectively educate patients for the self-management of their chronic conditions [24] and promote adherence to medical regimens [1].

For instance, the study by DeShazo et al. (2010) [10] demonstrated that serious games could significantly improve diabetes self-management among patients by enhancing their understanding and engagement with their treatment plans. Another study by Cafazzo et al. (2012) [5] found that a mobile-based serious game improved medication adherence among adolescents with diabetes by making the management of their condition more engaging and less burdensome.

Benchmarking Top Serious Games in Healthcare. Serious games have emerged as powerful tools for improving healthcare outcomes, particularly in enhancing patient adherence to treatment and fostering positive behavior change. Leading examples include Re-Mission, which helps cancer patients understand and manage their treatment [14]; EndeavorRx, the first FDA-approved game for ADHD treatment in children [2]; MySugr, a gamified diabetes management app that helps users track blood sugar and set personalized goals [15] and SuperBetter, which strengthens mental resilience and supports those managing anxiety and depression [16]. These platforms leverage game dynamics, reward systems, and educational content to engage users and promote healthier behaviors.

CuidaMED's AI-Driven Innovation. Where CuidaMED differentiates itself is through its integration of gamification with generative AI, setting a new benchmark in serious games for healthcare. CuidaMED utilizes ChatGPT, a generative AI that automates the creation of content, allowing for rapid and personalized game development. This AI-driven approach powers everything from storyline design to image generation, enabling the swift production of tailored game experiences that adapt to patient preferences. Unlike traditional serious games that rely on static narratives, CuidaMED dynamically adjusts to various patient needs, offering interactive narratives across genres such as suspense, science fiction, drama, or even real-life scenarios.

In addition, CuidaMED includes a virtual companion, Mary, who provides real-time feedback and rewards based on patient progress. This personalized interaction not only reinforces positive behaviors but also keeps patients engaged over time. Moreover, caregivers and family members can receive updates on the patient's condition, filling a critical gap in current healthcare gamification solutions. CuidaMED's use of generative AI thus revolutionizes the interaction between patients and digital healthcare tools, making chronic disease management more dynamic, adaptive, and engaging than ever before.

By benchmarking against leading serious games and incorporating cutting-edge AI technology, CuidaMED demonstrates how gamification in healthcare can be further industrialized and personalized. This approach sets a new standard for improving patient outcomes in chronic disease management, offering a scalable and adaptable solution that can evolve with patient needs.

3.5 CuidaMED Gamified Design

CuidaMED is a mobile application designed to help patients comply with their treatments. It allows patients to set schedules, report medical indicators, diet, exercise, and receive reminders. CuidaMED tracks progress and rewards patients with points that can be exchanged for gift cards and discounts. The app uses storytelling, inviting patients to engage in interactive games and make choices that earn points.

Additionally, CuidaMED includes serious games as an innovative feature to motivate patients to learn more about their conditions. These serious games use interactive scenarios to educate patients about their disease, treatment options, and the importance of adherence. By integrating serious games, CuidaMED leverages the educational potential of gamification to enhance patient engagement and self-management.

Octalysis Framework and Gamification in CuidaMED. CuidaMED's gamification approach is designed using the Octalysis framework, a systematic method for linking game mechanics to human motivation. The framework is based on eight core drives that influence behavior, and CuidaMED incorporates these principles to motivate users to adhere to their treatments and actively participate in their healthcare journey. Below is an explanation of how CuidaMED applies each of these core drives to engage and motivate its users.

- **Epic Meaning and Calling:** From the first day of using CuidaMED, users are welcomed with special rewards, such as free access to exclusive games. This makes the experience feel significant and makes users feel valued. The virtual companion, Mary, continually reinforces the user's importance by emphasizing that adhering to their treatments can prevent complications and save lives, creating a deep sense of purpose.
- **Development & Accomplishment:** CuidaMED users are driven by a sense of progress toward their health goals. The app provides a real-time dashboard displaying a graphical representation of completed and missed activities, adherence levels, and total points earned. This clear visualization of their journey empowers users to take control of their health, stay on track, and make informed decisions to maintain adherence.
- **Empowerment of Creativity & Feedback:** Users are motivated by the excitement of unlocking new milestones after completing activities or finishing games. CuidaMED introduces progressively more challenging games, rewards, and opportunities as users master their treatment plans, providing continuous feedback and keeping the experience fresh and rewarding.
- **Ownership & Possession:** CuidaMED fosters a sense of ownership by encouraging users to invest time in logging their prescribed activities and tracking their progress. The app also tailors the experience to each individual by offering personalized messages from Mary based on the user's progress, fostering an intimate connection between the user and the app.
- **Social Influence & Relatedness:** CuidaMED taps into social motivation by allowing family members and friends to send congratulatory messages when a user reaches significant milestones, such as a week of full adherence, or express concern when the user deviates from their treatment plan. This positive social reinforcement helps to strengthen adherence and engagement.

- **Scarcity & Impatience:** To incentivize users to stick to their treatments, CuidaMED offers special rewards that can be unlocked for completing a full week of adherence. These time-sensitive rewards create a sense of urgency and encourage users to consistently complete their tasks, reinforcing the idea that their time and effort in adhering to their treatment plan is valuable.
- **Unpredictability & Curiosity:** CuidaMED keeps the experience exciting by surprising users with random rewards, such as special gifts or the release of new games during notable events like birthdays. This unpredictability sparks curiosity and keeps users returning to the app, eager to discover what comes next.
- **Loss & Avoidance:** CuidaMED also motivates users by encouraging them to avoid missing out on rewards. For example, the app offers exclusive bonuses such as double points for maintaining a streak, and Mary keeps the user informed of these opportunities, ensuring that users are aware of the rewards they could lose if they fail to adhere to their treatments.

  By applying the Octalysis framework, CuidaMED delivers a comprehensive and engaging gamified experience that taps into multiple motivational drivers. This innovative approach ensures that patients are not only educated about their health, but also remain motivated to actively manage their conditions and adhere to their treatments, making CuidaMED a powerful tool for chronic disease management, show Fig. 1.
- **User Interface**

  CuidaMED leverages behavioral design and gamification to support chronic disease patients in managing their health. The app's interface combines simplicity with engaging elements that make the health management process rewarding.

  At the core of CuidaMED's interface is a real-time dashboard that displays users' adherence to treatments, physical activities, and dietary tracking. This dashboard is cleanly designed with visual cues like progress bars and color-coded indicators, helping users see their achievements and pending tasks at a glance.

  The app integrates artificial intelligence and a virtual caregiver, Mary, who interacts with users through friendly notifications and reminders. Mary provides tailored messages, offers feedback based on progress, and informs patients of upcoming healthcare appointments or medication schedules.

  A key feature is real-time tracking, allowing users to log daily adherence to their treatment plan. Each completed task earns points, while missed tasks are logged to provide a clear view of patient behavior.

  To motivate physical activity and healthy eating, personalized alerts are triggered, keeping users on track with their healthcare regimen. The app also incorporates interactive games that engage patients in learning about their disease management. These games combine storytelling and decision-making tasks, with rewards tied to user choices.

  Patients accumulate points through consistent adherence and progress, which can be redeemed for real-life rewards such as gift cards and discounts. This

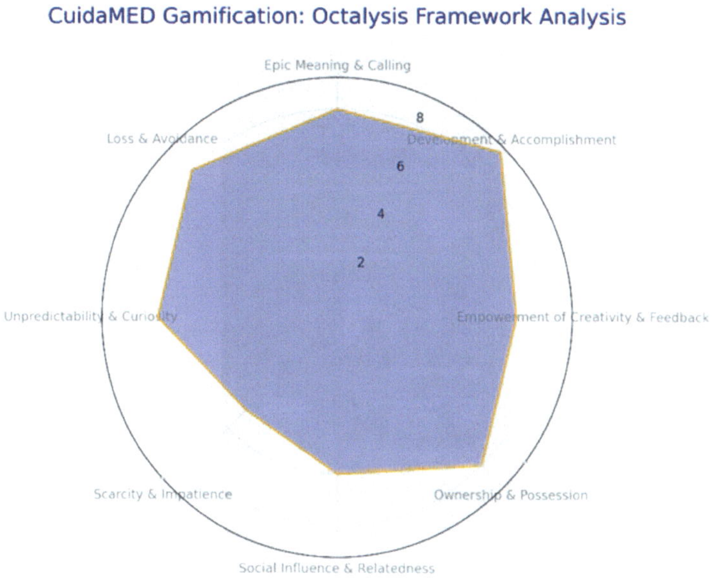

Fig. 1. Octalysis framework for the CuidaMED app. Each core drive is rated based on how CuidaMED utilizes it in its gamified design, ranging from Epic Meaning & Calling to Loss & Avoidance. The scores reflect the strength of each core drive in motivating users to adhere to their treatments and engage with the app. Representation based on The Octalysis Framework [8].

rewards system adds a layer of motivation, making the healthcare journey more engaging, effective, and enjoyable.

Below Figs. 2 and 3 are some screen shots from the CuidaMED app:

Gamification and serious games offer promising approaches to improving patient adherence and engagement in chronic disease management. In fact, gamification and serious games can improve patient education, understanding of their condition, and sense of ownership over their treatment, ultimately enhancing motivation and compliance.

4 Software Architecture and Interoperability

In this section, we explore the technological architecture underpinning CuidaMED, our digital tool designed to enhance patient adherence using behavioral science techniques. This architecture is built on the Fast Healthcare Interoperability Resources (FHIR) standard for health record management and integrates with clinic health information system (HIS) applications, along with a SMART on FHIR (Substitutable Medical Applications, Reusable Technologies on Fast Healthcare Interoperability Resources) app deployed on Azure.

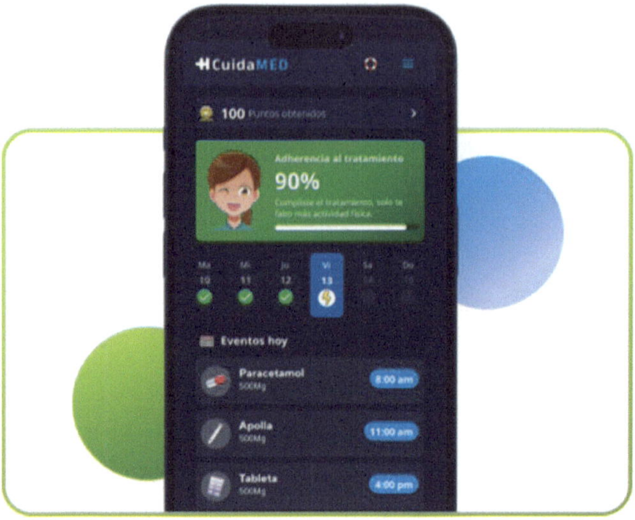

Fig. 2. CuidaMED dashboard features Mary, the virtual companion (Source: www.techealthlab.com)

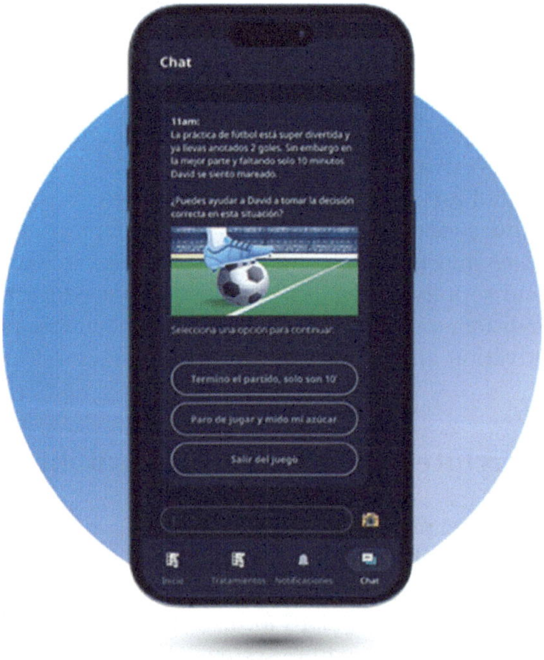

Fig. 3. CuidaMED serious game (Source: www.techealthlab.com)

CuidaMED stands out by incorporating the latest AI interaction patterns, specifically large language models (LLMs), to understand and respond to patient needs through natural human language.

The architecture comprises several layers, each serving specific functions to ensure the interoperability, security, and efficacy of the solution (show Fig. 4).

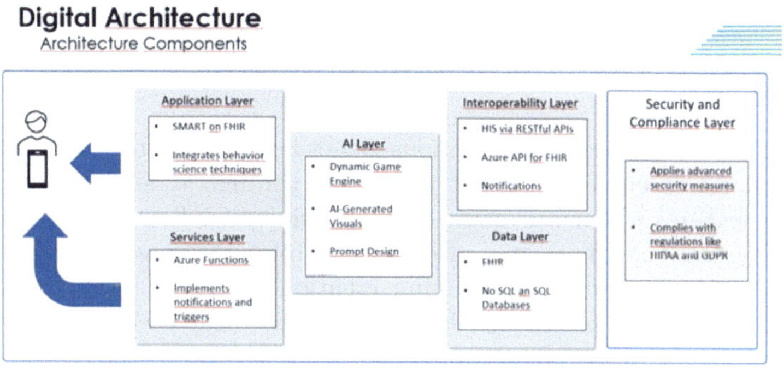

Fig. 4. CuidaMED Digital Architecture

4.1 Data Layer

The data layer relies on FHIR resources to standardize and structure clinical data, ensuring interoperability among various HISs. This enables CuidaMED to access and utilize consistent, up-to-date patient data and treatment information, facilitating personalized interventions based on user behavior.

4.2 Interoperability Layer

Interoperability is critical for real-time data exchange among different HISs and health applications. We implemented an FHIR server (e.g., Azure API for FHIR) handling FHIR-based requests and responses. This effectively integrates data from clinics' legacy systems using interoperable standards, fostering adoption and continuous use of CuidaMED by healthcare professionals and patients.

4.3 Application Layer

At the application layer, we developed the CuidaMED SMART on FHIR app designed to improve patient adherence. By integrating behavioral science techniques such as habit loops and reward systems, the app provides motivation and facilitates user action. This aligns with the Fogg behavior model, where patient motivation and ability to perform actions are crucial for effective behavior change.

4.4 Services Layer

We leverage Azure cloud services to implement business logic and advanced functionalities such as predictive analytics and user experience personalization. These services enhance user capability to perform specific actions, increase motivation through personalized notifications and reminders, and provide triggers based on clinical and behavioral data. These elements are essential per the Fogg behavior model to effectively trigger desirable behaviors.

4.5 AI Layer

In CuidaMED's architecture, the serious games layer integrates human language, LLMs, and AI to enhance patient adherence through interactive experiences. This layer features a dynamic game engine designed for flexible storytelling and a reward system that motivates progress and reinforces positive health behaviors. Generative AI streamlines content creation by automating storylines and adapting narratives to patient needs, ensuring alignment with therapeutic goals.

AI-generated visuals enhance immersion by reflecting patient choices in real-time. The layer also incorporates real-time feedback from patients and healthcare professionals, allowing the game to evolve dynamically. The design of the prompt for Mary, the virtual companion, is crucial as it defines her role in interacting with users, providing feedback, and connecting with health indicators and rewards, ensuring cohesive integration within the overall game architecture.

4.6 Load and Stress Testing

To validate the effectiveness and scalability of CuidaMED, comprehensive load and stress tests were conducted. These tests assessed the system's ability to handle large volumes of data and simultaneous users, ensuring the application can perform optimally under intensive usage scenarios. The results of these tests support our ability to deliver a robust and reliable solution that meets the performance requirements necessary to effectively enhance patient adherence and health outcomes.

For these tests, we utilized the Jmeter tool. Using this tool, we successfully emulated the behavior of 50 patients across all defined use cases in the CuidaMED app, reaffirming that our implemented technical capabilities allow for both horizontal and vertical scalability without any modifications.

During this monitoring period, we evaluated all defined use cases within the project. Figure 5 is a summary of the major epics monitored:

We monitored all functionalities implemented in the CuidaMED app, covering a range of health management tasks. During this process, we conducted tests with 50 concurrently active patients to evaluate the performance of each application functionality.

The 50 patients were brought online with a ramp-up curve of 250 s, equating to one new patient entering every 5 s (show Table 2).

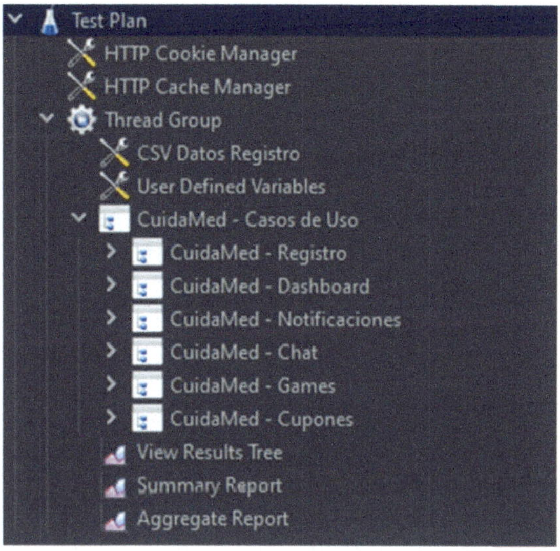

Fig. 5. CuidaMED Epics Monitored

Table 2. Table of Monitoring Results

Heading level	Example	Average	Error %
Registration	50	412	0.00%
Dashboard	50	237	0.00%
Notifications	50	19192	0.00%
Chat	50	608	0.00%
Games	50	315	0.00%
Rewards	50	3813	0.00%

The load test involving 50 patients showed no errors, indicating that the implemented technical capabilities can support a significant patient load simultaneously without any adjustments to the current architecture. This ensures that the application can scale exponentially without major changes.

The most important findings from the conducted monitoring are as follows:

a) Consistent throughput: Overall, there is good performance in terms of throughput for all transactions, indicating the system's ability to handle concurrent user loads effectively.
b) Acceptable response times: Most transactions exhibit acceptable average response times, suggesting efficient request processing.
c) Low error rate: A recorded error rate of 0.00% reflects remarkable system stability during testing, with no significant failures.
d) Good performance in specific queries**: All transaction groups demonstrate satisfactory response times and throughput.

In summary, CuidaMED's multi-layered architecture, combined with generative AI-powered serious games, offers a robust, scalable, and secure solution for enhancing patient adherence in chronic disease management. The AI-driven game design accelerates development, adapts content to individual patients, and incorporates real-time feedback, making CuidaMED a highly responsive and engaging tool in healthcare. The system's load-testing results validate its capacity to serve large patient populations while maintaining high levels of performance, positioning CuidaMED as an innovative and effective platform for improving patient outcomes.

5 Effectiveness of CuidaMED

5.1 Peruvian HealthCare

In the broader context of the healthcare market in Latin America, particularly in Peru, understanding the reality of these markets is crucial for interpreting the results obtained from testing the CuidaMED tool.

According to the National Institute of Statistics and Informatics (INEI)[13], the average waiting time for care at MINSA establishments is 114 min, at EsSALUD 58 min, at the Armed Forces and Police 50 min, and at private clinics 35 min. Research found that 51.2% of IPRESS users did not receive information about waiting times. Peru also has a 10.5% adverse event rate during hospitalizations, with 7% mortality and varying degrees of disability.

Recent data underscores the critical role of pharmacies in Peru's healthcare system. On September 9, 2023, INEI reported that 17.4% of individuals with health issues sought care at pharmacies, with urban areas showing 17.6% and rural areas 16.4%. Additionally, on March 21, 2024, INEI noted that 21.4% of individuals sought care at pharmacies, up from 20.5% the previous year. The June 28, 2024 ENAHO report highlighted that 11.6% of people aged 60+ sought attention at pharmacies in the first quarter of 2024, indicating a growing reliance on these facilities.

These figures illustrate the increasing dependence on pharmacies for healthcare, revealing gaps in the public system that the platform aims to address.

To evaluate the effectiveness of the tool addressing this issue, we monitored a focused group of patients (F&F) and conducted interviews with a larger group to gather insights into their healthcare experiences and the technological solutions they use to enhance adherence and achieve better health outcomes. The interviews were complemented by detailed presentations on the CuidaMED app's scope, analyzing how this application could benefit patients in maintaining adherence and promoting a healthy lifestyle. This report focuses on the conclusions derived from two phases of patient adherence monitoring during these tests. Patient identities have been protected to ensure data security and confidentiality.

5.2 Focused Group of Patients (F&F) Results

The first stage of the evaluation involved monitoring 14 patients with chronic conditions from the F&F circle over a month to assess the impact of using the

CuidaMED app on their adherence to treatment. Results showed significant changes in adherence after the introduction of the app (show Table 3):

Table 3. Significant Changes in Adherence After App Introduction

Patient	Condition	Pre-Adherence	Post-Adherence	Improvement
1	Thyroid	79%	90%	+11%
2	Hypertension + Hypertension	79%	90%	+11%
3	Hyperthyroidism + Diabetes	71%	90%	+19%
4	High Blood Pressure	93%	93%	0%
5	Diabetes	65%	80%	+15%
6	Thyroid	78%	88%	+10%
7	Hypertension	85%	89%	+4%
8	Hyperthyroidism	70%	84%	+14%
9	High Blood Pressure	90%	91%	+1%
10	Diabetes + Hypertension	68%	82%	+14%
11	Thyroid + Diabetes	72%	86%	+14%
12	Hypertension + Hyperthyroidism	80%	85%	+5%
13	Diabetes + High Blood Pressure	77%	88%	+11%
14	Hypertension + Diabetes	74%	87%	+13%

The second stage involved interviewing 20 patients with chronic conditions. A set of more than 20 questions was designed, including several open-ended questions to allow patients to freely share their opinions without bias. They were also given a live demonstration of the CuidaMED app and allowed to navigate its features, simulating their interaction with the application.

The general conclusion was that 50% of the patients indicated that the app would improve their adherence by at least 30%, while the remaining patients felt the app was very convenient and would improve their adherence by 15% to 20%. These figures align with those observed in the first stage of the adherence improvement evaluation.

Key highlights from patient feedback included:

- **Gamification:** Patients found that gamification elements, such as rewards and challenges, made tracking their treatment more engaging and motivating.
- **Interactive Reminders:** Personalized and interactive reminders for medication and treatments were seen as more effective than traditional methods.
- **Real-Time Monitoring:** Patients appreciated the potential for immediate intervention in case of deviations or emergencies.
- **Continuous Education:** Ongoing educational information was valued for improving understanding and long-term benefits of treatment.

- **Community and Social Support:** Integration of social functions and the ability to connect with family and friends created a sense of community and support.
- **Playful Elements:** Game-like elements such as missions and levels made treatment tracking a more positive experience.
- **Progress Visualization:** Visual graphs of progress were seen as an additional motivation to stay on track.
- **AI and LLM Integration:** Personalized reminders and educational content, and trigger alerts for timely interventions. AI adapts to patient profiles, enhancing personalization and engagement.

6 Security and Privacy

We ensure that the architecture complies with rigorous security and privacy standards such as the Health Insurance Portability and Accountability Act (HIPAA) and General Data Protection Regulation (GDPR), leveraging advanced Azure security tools. These tools are pivotal in protecting sensitive patient data against unauthorized access, breaches, and ensuring compliance with regulatory frameworks specific to healthcare. By adopting Azure's comprehensive security measures, including encryption, access controls, audit logs, and continuous monitoring, our solution not only meets legal requirements but also establishes a foundation of trust with users of CuidaMED.

Azure's security and compliance blueprint for HIPAA/HITRUST provides a detailed framework and best practices for implementing robust security controls in healthcare environments [17]. This blueprint aligns with industry standards and guidelines, offering a structured approach to safeguarding health data and maintaining compliance. By adhering to these standards, we enhance user confidence in CuidaMED's security posture, promoting greater patient engagement and sustained use of the application. This, in turn, supports our goal of improving patient outcomes and adherence through secure and reliable health management solutions.

7 Conclusions

The development and implementation of CuidaMED mark a significant advancement in addressing the pervasive issue of patient non-adherence in chronic disease management. Our research underscores the efficacy of integrating behavioral economics and gamification principles within healthcare applications to foster long-term adherence and improve health outcomes. Below are the key findings and implications from the research:

Improved Patient Engagement: Our pilot studies and usability testing revealed that CuidaMED successfully enhances patient engagement, showing potential improvement in adherence to treatment regimens. By incorporating a virtual companion, Mary, and gamified elements such as interactive games and a point-based reward system, the application transforms the challenging task of managing chronic conditions into an engaging and motivating experience. The positive feedback from participants indicates that the gamification and serious games elements effectively sustain patient motivation and interest over time.

Behavioral Economics and Habit Formation: The application of behavioral economics theories, such as the Fogg behavior model and the Octalysis framework, proved instrumental in designing interventions that align with actual human behavior. By addressing common barriers to adherence, such as present bias and the intent-behavior gap, CuidaMED effectively incentivizes patients to follow their treatment plans. The incorporation of habit formation strategies, such as the habit loop theory, further reinforces the development of sustainable healthy behaviors.

Technological Innovation and Interoperability: CuidaMED's robust technological architecture, built on FHIR standards and integrated with clinic HIS applications, ensures seamless interoperability and real-time data exchange. This integration is crucial for providing personalized interventions and real-time monitoring of patient progress. The scalability and reliability of the system, validated through comprehensive load and stress testing, affirm CuidaMED's capability to handle significant patient loads without compromising performance. Additionally, the use of generative and human language AI in the platform has accelerated and industrialized the design and construction of serious games, allowing for faster development and more efficient scaling of game elements. Moreover, generative AI enables dynamic adaptation of content based on different population segments, tailoring games to the specific needs and characteristics of patients, thus enhancing engagement and adherence.

Security and Privacy: Adhering to stringent security and privacy standards, such as HIPAA and GDPR, CuidaMED ensures the protection of sensitive patient data. The adoption of advanced Azure security tools provides a secure environment for data storage and exchange, fostering trust and confidence among users. This compliance with regulatory frameworks is essential for the sustained use and success of the application.

Social and Economic Impact: Further research is needed to assess the long-term effects of CuidaMED on patient adherence and health outcomes.

The successful results from the tests and interviews highlight the significant impact that CuidaMED can have on patients by fostering adherence through gamification and engaging experiences. This is especially relevant in the Peruvian healthcare context, where challenges such as long wait times, lack of transparency, high prevalence of adverse events, and short consultation times contribute to dissatisfaction and lower adherence rates.

- Wait times in health facilities: High average wait times can negatively impact patient adherence by causing frustration and demotivation.

- Information on wait times: The lack of clear information on wait times contributes to dissatisfaction and inefficiency in healthcare.
- Prevalence of adverse events: The high rate of adverse events during hospitalizations underlines potential deficiencies in care quality, which can erode trust and affect adherence.
- External consultation times: Short consultation times may contribute to a poor overall experience and low adherence rates.

Future studies should involve larger, more diverse populations to validate the findings and explore advanced AI-driven personalization and analytics to enhance the platform's efficacy.

Future Directions: While the initial results are promising, further research is needed to explore the long-term effects of CuidaMED on patient adherence and health outcomes. Future studies should include larger and more diverse populations to validate the findings and refine the application's features. Additionally, exploring the integration of more advanced AI-driven personalization and real-time analytics could further enhance the application's efficacy.

In conclusion, CuidaMED represents a groundbreaking solution to the critical issue of patient non-adherence in chronic disease management. By leveraging behavioral economics, gamification, and cutting-edge generative AI technology, the platform not only motivates patients to adhere to their treatments but also accelerates the design, industrialization, and adaptation of interactive games based on population needs. CuidaMED's robust technological architecture ensures seamless integration, scalability, and security, making it a scalable and effective tool for improving health outcomes. This study emphasizes the transformative potential of innovative healthcare applications and generative AI, offering a promising path forward in addressing chronic disease management challenges.

References

1. Abraham, O., LeMay, S., Bittner, S., Thakur, T., Stafford, H., Brown, R.: Investigating serious games that incorporate medication use for patients: Systematic literature review. JMIR Serious Games **8**(2), e16096 (2020). https://doi.org/10.2196/16096
2. Akili Interactive: Akili announces endeavor TM attention treatment is now available for children with attention deficit hyperactivity disorder (ADHD) (2020). https://www.akiliinteractive.com/news-collection/akili-announces-endeavortm-attention-treatment-is-now-available-for-children-with-attention-deficit-hyperactivity-disorder-adhd-al3pw
3. Ariely, D.: Predictably irrational: the hidden forces that shape our decisions. cj Advertising, Ebook, Revised (1998)
4. Brown, M.T., Bussell, J.K.: Medication adherence: who cares? Mayo Clin. Proc. **86**(4), 304–314 (2011). https://doi.org/10.4065/mcp.2010.0575. Epub 2011 Mar 9

5. Cafazzo, J.A., Casselman, M., Hamming, N., Katzman, D.K., Palmert, M.R.: Design of an mhealth app for the self-management of adolescent type 1 diabetes: a pilot study. J. Med. Internet Res. **14**(3), e70 (2012). https://doi.org/10.2196/jmir.2058
6. Charness, G., Gneezy, U.: Incentives to exercise. Econometrica **77**(3), 909–931 (2009)
7. Chou, Y.K.: Actionable Gamification: Beyond Points, Badges, and Leaderboards. Packt Publishing Ltd. (2019)
8. Chou, Y.K.: Octalysis: complete gamification framework (2020). https://yukaichou.com/gamificationexamples/octalysis-complete-gamification-framework/
9. Clear, J.: Atomic Habits: An Easy & Proven Way to Build Good Habits & Break Bad Ones. Penguin, USA (2018)
10. DeShazo, J., Harris, L., Pratt, W.: Effective intervention or child's play? A review of video games for diabetes education. Diab. Technol. Ther. **12**(10), 815–822 (2010). https://doi.org/10.1089/dia.2010.0030
11. Duhigg, C.: The Power of Habit: Why We Do What We Do in Life and Business. Random House (2012)
12. Fogg, B.: A behavior model for persuasive design. In: Proceedings of the 4th International Conference on Persuasive Technology, Persuasive '09. Association for Computing Machinery, New York, NY, USA (2009). https://doi.org/10.1145/1541948.1541999
13. Instituto Nacional de Estadística e Informática (INEI): Usuarios esperan más de una hora para ser atendidos en un establecimiento de salud. https://m.inei.gob.pe/prensa/noticias/usuarios-esperan-mas-de-una-hora-para-ser-atendidos-en-un-establecimiento-de-salud-7823/. Accessed 27 July 2024
14. Kato, P.M., Cole, S.W., Bradlyn, A.S., Pollock, B.H.: A video game improves behavioral outcomes in adolescents and young adults with cancer: a randomized trial. Pediatrics **122**(2), e305–e317 (2008). https://doi.org/10.1542/peds.2007-3134
15. Krebs, P., Duncan, D.T.: Health app use among us mobile phone owners: a national survey. JMIR Mhealth Uhealth **3**(4), e101 (2015). https://doi.org/10.2196/mhealth.4924
16. McGonigal, J.: SuperBetter: A Revolutionary Approach to Getting Stronger, Happier, Braver and More Resilient. Penguin, USA (2015)
17. Microsoft: Microsoft's security and compliance blueprint for HIPAA/HITRUST health data & AI (2024). https://learn.microsoft.com/en-us/azure/architecture/solution-ideas/articles/security-compliance-blueprint-hipaa-hitrust-health-data-ai
18. Osterberg, L., Blaschke, T.: Adherence to medication. N. Engl. J. Med. **353**(5), 487–497 (2005). https://doi.org/10.1056/NEJMra050100
19. Petry, N.M., Alessi, S.M., Byrne, S., White, W.B.: Reinforcing adherence to antihypertensive medications. J. Clin. Hypertens. **17**(1), 33–38 (2015). https://doi.org/10.1111/jch.12441
20. Sabaté, E. (ed.): Adherence to Long-term Therapies: Evidence for Action. World Health Organization (2003)
21. Sen, A.P., et al.: Financial incentives for home-based health monitoring: a randomized controlled trial. J. Gen. Intern. Med. **29**(5), 770–777 (2014). https://doi.org/10.1007/s11606-014-2778-0

22. Superintendencia Nacional de Salud (SUSALUD): Resultados definitivos en susalud 2016 (2016). https://portal.susalud.gob.pe/wp-content/uploads/archivo/encuesta-sat-nac/2016/PRESENTACION-SUSALUD-2016.pdf
23. Thaler, R.H., Sunstein, C.R.: Nudge: Improving Decisions about Health, Wealth, and Happiness. Yale University Press (2008)
24. Thomas, T.H., Sivakumar, V., Babichenko, D., Grieve, V.L.B., Klem, M.L.: Mapping behavioral health serious game interventions for adults with chronic illness: scoping review. JMIR Serious Games **8**(3), e18687 (2020). https://doi.org/10.2196/18687

AI and Data Analytics

Voice-Command Responsive Autonomous Navigation for Service Robots Using the Ollama Framework on the ARMOS TurtleBot

Fredy Martínez(✉) and Angélica Rendón

Universidad Distrital Francisco José de Caldas, Bogotá D.C., Colombia
fhmartinezs@udistrital.edu.co

Abstract. Autonomous navigation in dynamic environments presents significant challenges in robotics, particularly in adapting to real-time verbal instructions. This paper introduces a novel approach using the ARMOS TurtleBot, integrated with the Ollama framework to process and execute voice commands. We developed a user-friendly voice interface allowing the robot to interpret commands such as "*move forward*," "*move backward*," "*turn left*," "*turn right*," and "*stop*" through natural language. Controlled tests evaluated the system's responsiveness and accuracy, demonstrating the robot's capability to adapt its path based on real-time verbal directions. This research highlights the integration of language processing technologies with autonomous robotics, enhancing robotic autonomy and offering practical solutions for service robots, assistive technologies, and industrial automation. The findings contribute significantly to the field, showcasing potential applications in dynamic interaction environments.

Keywords: autonomous navigation · dynamic environments · language processing · real-time interaction · service robotics · voice control

1 Introduction

Autonomous navigation has emerged as an essential technology in robotics [31], finding diverse applications across multiple industries, including personal assistance, logistics, and manufacturing [1]. In personal assistance, robots simplify daily tasks for the elderly or disabled, enhancing their quality of life by enabling greater independence [29]. In logistics, autonomous vehicles streamline the complexities of warehouse management, optimizing the supply chain by automating goods transportation and sorting processes [11]. Similarly, in manufacturing, robots equipped with autonomous navigation capabilities are instrumental in carrying out intricate tasks such as assembling, painting, and inspection, which

F. Suni-Lopez et al. (Eds.): CIIS 2024, ISEM 48, pp. 53–67, 2025.
https://doi.org/10.1007/978-3-031-91481-2_4

are not only monotonous but also potentially hazardous to humans [10]. The integration of autonomous navigation systems in these sectors not only increases efficiency, but also reduces operational costs and human error, significantly impacting productivity and safety.

However, deploying autonomous robots in dynamic environments poses significant challenges [12,18]. Unlike controlled environments where variables are predictable and static, dynamic settings are characterized by their unpredictability and constant change, presenting unique obstacles and situations that require immediate and accurate responses [20]. For instance, a service robot navigating a busy hospital corridor must be capable of real-time decision-making to avoid collisions with moving obstacles like people, carts, and other robots [26]. Moreover, the ability to understand and adapt to verbal instructions in such environments enhances a robot's flexibility and utility, making it more suitable for real-world applications where conditions and requirements can change rapidly [2]. This necessity for high adaptability and precise real-time processing in autonomous navigation highlights the critical need for advanced solutions that can effectively interpret human commands and translate them into actionable tasks, ensuring both operational efficacy and safety [15].

While autonomous navigation technologies have significantly advanced, traditional navigation systems often exhibit notable limitations, particularly in terms of flexibility and real-time adaptability [6]. These systems are generally designed for environments with predictable parameters, where changes occur over more extended periods, allowing for pre-scheduled updates and adjustments. However, in dynamic settings such as public spaces or emergency scenarios, these systems struggle because they cannot quickly recalibrate to sudden environmental changes or unpredicted obstacles [14,30]. This rigidity can lead to inefficiencies or even failures in task execution, compromising the robot's usefulness in situations that demand agile and immediate decision-making.

Moreover, a critical gap in current robotics technology is the interface for human-robot interaction, which has not evolved at the same pace as other technological aspects [9]. Most traditional robotic systems rely on pre-programmed commands or interface through limited touch or text-based inputs, which do not mimic natural human communication [23]. The ability to process and respond to spoken instructions remains underdeveloped, which restricts the deployment of robots in roles that require direct interaction with untrained users [32]. Enhancing voice command capabilities and developing more intuitive interaction methods are essential for expanding the applicability of robotics into new areas, such as personalized customer service or support for individuals with disabilities [4,19]. Addressing these deficiencies by integrating sophisticated language processing algorithms like those provided by Ollama could revolutionize how robots understand and execute tasks, making them far more adaptable and user-friendly in everyday applications.

To address the articulated deficiencies in robotic navigation systems, the integration of Ollama offers a promising avenue [27]. Ollama is a sophisticated framework capable of efficiently processing natural language and executing language models, which positions it as an ideal solution for enhancing human-robot interaction [7]. Its ability to interpret and respond to verbal commands in real time can significantly mitigate the rigidity of traditional navigation systems, allowing robots to adapt swiftly to unanticipated environmental changes and instructions [21]. This adaptability is crucial in dynamic settings where traditional systems falter, due to their inability to rapidly incorporate new data and adjust their operations accordingly [3]. By leveraging Ollama's advanced language processing capabilities, robots can achieve a higher degree of autonomy and flexibility, making them more effective and responsive in environments that closely mimic human operational contexts [13]. Furthermore, the integration of Ollama enhances the interaction layer of robotics, providing a more natural and intuitive user experience, which is vital for applications involving direct human contact, such as in healthcare, customer service, and personal assistance. This advancement not only bridges the interaction gap, but also extends the potential applications of robotics into sectors where communication fluidity and understanding are paramount.

The primary objective of this study is to develop a system that enables the ARMOS TurtleBot to navigate autonomously while responding effectively to voice commands [16,17]. This system aims to leverage the capabilities of the Ollama framework to process natural language in real-time, thereby allowing the robot to interpret and act upon verbal instructions seamlessly. The integration of Ollama is expected to significantly enhance the robot's flexibility and utility in dynamic environments, where rapid adaptation to new instructions and obstacles is crucial. This research intends to demonstrate that enhanced voice command capabilities can bridge the current gap in human-robot interaction, leading to more intuitive and effective communication methods.

The expected contributions of this research are manifold. By enabling the ARMOS TurtleBot to navigate autonomously based on voice commands, we anticipate improvements in the interaction between humans and robots, making such systems more accessible and user-friendly. Additionally, this capability is likely to enhance the efficiency and safety of autonomous navigation, particularly in complex, unpredictable settings where traditional navigation systems fall short. Ultimately, this study aims to show that integrating advanced language processing with robotic systems can expand their applicability across various sectors, potentially transforming how robots are utilized in everyday settings and specialized applications alike. These advancements could lead to broader adoption of robotic solutions in industries where dynamic interaction and adaptability are valued, further pushing the boundaries of what autonomous systems can achieve.

The remainder of this paper is structured as follows: Sect. 2 examines prior research on autonomous navigation and human-robot interaction. Section 3 describes the integration of the Ollama framework with ARMOS TurtleBot and the experimental setup. Section 4 presents our findings on the system's performance in dynamic environments. Finally, Sect. 5 summarizes the study's outcomes, discusses the implications for future robotic applications, and suggests directions for further research.

2 Literature Review

The exploration of autonomous navigation systems for service robots has garnered substantial attention within the robotics community, spurred by the need for robust solutions in dynamic environments. Chung [8] pioneered an integrated navigation system for indoor service robots, capable of operating in extensive environments and performing a range of tasks including delivery and patrol. This work laid foundational principles for subsequent studies, emphasizing the importance of scalability and versatility in robot navigation systems. Following this, Park [22] introduced an intelligent navigation method that utilizes semantic maps to facilitate path planning in complex environments, thereby enhancing the interaction between robots and their operational spaces. Additionally, Ragavan [25] developed a telematics framework tailored for autonomous robots, integrating communication, decision support, and sensor fusion modules to address navigation and sensory challenges in real-time operations.

Further advancements were made in the software frameworks supporting mobile robotics, as demonstrated by Beck [5], who introduced MobotWare, a modular, plug-in-based software system designed to support both educational and research applications in mobile robotics. This framework not only facilitates the integration of various robotic functionalities, but also promotes the adaptability and expandability of robotic systems. In parallel, the realm of human-robot interaction has seen significant innovations aimed at enhancing safety and social compliance. For instance, Qian [24] proposed novel pre-collision safety strategies that prioritize human safety and comfort, ensuring that navigation strategies respect social norms and personal boundaries. These strategies are crucial in environments where robots and humans coexist closely, necessitating navigation systems that are both efficient and socially aware.

Recent studies, such as those by Truong [28], have further pushed the boundaries by focusing on socially aware navigation, which incorporates human detection and tracking to facilitate interactions that are deemed socially acceptable. This research underscores the growing necessity for service robots to be aware of their social surroundings, enabling them to operate seamlessly and unobtrusively in human-centric environments. The cumulative body of work highlights a clear trajectory towards more intelligent, adaptable, and socially integrated autonomous navigation systems, pointing towards a future where robots can perform more autonomously and effectively across various service tasks. Collectively, these studies lay a robust foundation for enhancing robotic capabilities, ensuring that service robots not only navigate efficiently, but also interact ethically and safely within their operational contexts.

3 Methods

3.1 Robot ARMOS TurtleBot

The ARMOS TurtleBot (Fig. 1), designed with a differential drive system, is equipped with four tracks, each powered by 9 V DC motors with a high starting torque of 9.5 kg and a no-load speed of 150 RPM. This setup ensures that the robot can maneuver effectively over various terrains, making it suitable for both indoor and outdoor environments. The motors are controlled by signals that simultaneously activate both motors on each side, simplifying the control scheme while enhancing the overall torque, crucial for overcoming obstacles and inclines.

Fig. 1. ARMOS TurtleBot, a robotic platform with displacement and loading capacity for assistive applications

In terms of sensory capabilities, the ARMOS TurtleBot is outfitted with an array of sensors to ensure robust environmental interaction and navigation precision. The primary navigation sensor is a LIDAR unit (RPLIDAR A1M8), which provides 360-degree coverage, crucial for obstacle detection and avoidance. This is supplemented by ultrasonic and infrared sensors that offer additional data points for closer range detection, thereby enhancing the robot's ability to navigate complex environments. The integration of these sensors is managed through a central processing unit that coordinates data collection and actuation, based on real-time analysis and pre-programmed algorithms.

Furthermore, the ARMOS TurtleBot is designed to be highly modular, allowing for easy customization and scalability. This modularity is evident in its ability to integrate additional sensors or actuators without requiring major hardware modifications. The robot's software is built on a robust operating system that supports real-time operations, ensuring reliable performance under varying operational conditions. This software architecture not only facilitates routine navigation tasks, but also supports complex interaction patterns, such as dynamic path planning and environment mapping, which are essential for autonomous operation in unpredictable settings.

3.2 Vosk-Based Voice Recognition

The integration of voice-driven interfaces into robotic systems presents a unique set of challenges and opportunities. To address this, we have implemented a voice recognition module using the Vosk API, a versatile speech-to-text library that supports multiple languages and operates efficiently under various acoustic conditions. This module is particularly tailored to enhance the interactive capabilities of the ARMOS TurtleBot, enabling it to understand and execute verbal commands through an integration with the Llama 3 language model.

The initial step in setting up the voice recognition system involved the installation and configuration of the Vosk library (Table 1). A suitable Vosk model was selected based on the operational language and the computational constraints of the host system. The chosen model was downloaded from the official Vosk model repository and integrated into our Python-based control system. This setup ensures that the speech recognition process is both accurate and efficient, providing a seamless interface for real-time robotic control.

The operations of the ARMOS TurtleBot were programmed in Python, utilizing the Robot Operating System for sensor integration and motion control. Robot Operating System provides a modular infrastructure that facilitates the development of complex robotic applications, including autonomous navigation and sensor management. Additionally, libraries such as Vosk for voice recognition and Ollama for natural language processing were employed, enabling seamless and real-time interaction between the robot and the user through voice commands.

To facilitate real time audio processing, we utilized the sounddevice library in Python, which allows for the streaming of microphone input directly into the Vosk recognition engine. The audio stream is captured at a sample rate of 16000 Hz, which is optimal for voice recognition. Each segment of the audio stream is processed in real-time, with Vosk converting spoken words into textual commands. This continuous stream processing is crucial for maintaining the responsiveness of the robot to dynamic commands in its operating environment.

The integration of the voice commands into the robotic control system is managed through a series of Python scripts that handle the interpretation and execution of commands. Once Vosk processes the audio input and extracts the command text, this text is passed to the Llama 3 model. Llama 3 then interprets the command within the context of the robot's current state and operational parameters. For instance, a command such as "*turn right*" is processed to understand the intent and the degree of rotation required based on the robot's current orientation and positional data.

To enable communication between the ARMOS TurtleBot and the processing device, a wireless connection via WiFi is utilized, allowing efficient data transmission without the need for cables that could limit the robot's mobility. The voice recognition system uses an omnidirectional microphone with noise-canceling capabilities, selected for its high sensitivity (frequency response of 20 Hz to 20 kHz) and its ability to capture commands in noisy environments. This choice is based on preliminary tests that showed a significant improvement

Table 1. Python framework showing how to set up voice command recognition with Vosk

```
import os
from vosk import Model, KaldiRecognizer
import sys
import json
import sounddevice as sd
import numpy as np

def int_or_str(text):
        """Helper function for argument parsing."""
        try:
        return int(text)
        except ValueError:
        return text

def callback(indata, frames, time, status):
        """Callback function to collect from microphone and send to Vosk"""
        if status:
        print(status, file=sys.stderr)
        if rec.AcceptWaveform(indata):
        result = json.loads(rec.Result())
        process_command(result['text'])

def process_command(command):
        """Process the command using Llama 3 and execute the action"""
        print(f"Received command: {command}")
        # Here, integrate with Llama 3 to interpret the command
        # This code assumes a function to handle the command interpretation
        # and execution
        execute_robot_command(command)

def execute_robot_command(command):
        """Placeholder function to execute commands on the robot."""
        # Convert command to actionable tasks for the robot
        print(f"Executing: {command}")
        # Example: If command is 'turn right', turn the robot to the right
        # Add control code here

# Load the Vosk model
model_path = "path/to/vosk/model"
model = Model(model_path)
rec = KaldiRecognizer(model, 16000)

# Set up audio stream
with sd.RawInputStream(samplerate=16000, blocksize = 8000, dtype='int16',
channels=1, callback=callback):
        print('#' * 80)
        print('Listening for commands...')
        print('#' * 80)
        while True:
                pass
```

in voice recognition accuracy by minimizing ambient noise interference. We used the Fifine K669B USB Microphone, which is widely available and offers a balanced combination of affordability and performance, making it suitable for real-time voice command applications.

3.3 Integration of Ollama

Ollama is a sophisticated software framework designed to efficiently process natural language and execute language models, enhancing human-machine interaction by enabling real-time interpretation and response to verbal commands. The integration of the Ollama platform, particularly the Llama 3 model, into our Linux Mint system (Kernel 6.9.6-1-liquorix-amd64) was initiated by a straightforward installation process. Llama 3 (Llama 3-8B.Q4) is an advanced large language model within the Ollama framework, known for its capability to process and generate natural language, facilitating complex and nuanced human-computer interactions in real-time.

```
curl -fsSL \url{https://ollama.com/install.sh} | sh
ollama run llama3
```

Using the command-line instruction, we installed Ollama on a system powered by an AMD Ryzen 7 2700X CPU and an NVIDIA GeForce GTX 1050 Ti GPU, with 16 GB of RAM. This hardware setup ensured adequate computational and graphical processing capabilities for handling the demands of natural language processing tasks.

After installation, we selected Llama 3 from the Ollama library, which is recognized for its ability to process and generate natural language processing effectively. This model was deemed appropriate for interpreting complex voice commands necessary for controlling a robot in dynamic environments. To train Llama 3, we prepared a CSV file named `training_data.csv` (Table 2). This file contained mappings of voice commands to specific robot actions, ensuring that each command such as "*turn right*" corresponded to a predefined action like "*DIRECTION: RIGHT*". The initial training table had 60 entries, but it is possible to continually update it with new expressions without modifying the architecture.

The model's performance was tested through a series of practical simulations. Commands such as "*make a complete rotation*" and "*can you go back?*" were issued to assess the model's ability to interpret and execute instructions accurately. Responses from Llama 3 were typically delivered within two to three seconds, demonstrating both the model's efficiency and its accuracy in command execution. For instance, the command "*make a complete rotation*" resulted in the correct simulation of a 360-degree turn by the robot.

This phase of integration highlighted Llama 3's capacity to accurately process voice commands and reliably perform corresponding actions. The setup proved effective in bridging the gap between natural language commands and robotic actions, crucial for enhancing the robot's performance in real-world tasks. The practical tests confirmed the model's suitability for deployment in dynamic settings where real-time command interpretation and execution are critical.

Although voice and natural language processing were conducted on a dedicated computer during the experiments, we explored the feasibility of running the libraries directly on a Raspberry Pi installed on the ARMOS TurtleBot. The Raspberry Pi could potentially handle the computational demands of the Llama

Table 2. Sample commands and corresponding actions from training data

Voice Command	Action
move forward	forward
go forward	forward
advance straight	forward
continue moving forward	forward
proceed straight	forward
march ahead	forward
advance	forward
step forward	forward
keep moving	forward
go ahead	forward
walk ahead	forward
move ahead	forward
proceed	forward
go back	backward
move backward	backward

3 model and the Vosk voice recognition; however, more robust hardware was chosen to ensure greater processing speed and flexibility during the experimental phase. Future implementations may consider this alternative to increase the system's autonomy.

3.4 Command Interpretation and Hardware Control Integration

This integration hinges on the efficient translation of voice commands, processed through Llama 3, into actionable hardware controls. Our implementation leverages Python's versatility and the strong interfacing capabilities provided by various libraries to facilitate communication between software commands and physical robot actions. The primary objective is to ensure that voice commands are not only accurately interpreted but are also effectively translated into the robot's motion and operations.

The initial step involves setting up a command interpretation mechanism. Utilizing Python, we developed a command controller that maps specific textual commands to corresponding robot actions. Each command processed by Llama 3, derived from user inputs captured and converted by Vosk, is associated with a predefined function within our control system. For instance, a command such as "*move forward*" is linked to a function that triggers the robot's motors to initiate forward movement. This mapping is stored in a Python dictionary, providing a quick lookup and execution path for each command.

Integration with the robot's hardware is facilitated through the PySerial library, which allows Python scripts to communicate with serial devices. In our setup, commands translated into serial commands are sent to the robot's microcontroller, which directly manages the actuators and motors based on the received instructions. The serial communication protocol was chosen for its reliability and ease of implementation on most Linux systems. The configuration of the serial connection, including baud rate and port settings, is tailored to match the specifications of the ARMOS TurtleBot's onboard communication module.

Furthermore, the implementation of the command controller includes error handling and feedback mechanisms to ensure smooth operation. Exception handling routines are integrated to manage potential issues such as communication timeouts or data corruption. Additionally, each command execution results in a confirmation message sent back from the robot's microcontroller, confirming the successful reception and execution of the command. This feedback is crucial for maintaining the reliability of the system, providing real-time monitoring capabilities to the user interface.

The user interface, developed using Python's Tkinter library, offers a graphical representation of the robot's status and available commands. This interface allows users to manually input commands or override automated functions, providing a direct control mechanism during testing and operation. The interface also displays system logs and feedback from the robot, assisting in diagnostics and monitoring during field operations.

Testing and validation of the system were conducted in a controlled laboratory environment, followed by field tests in a simulated real-world setting (Fig. 2). These tests assessed the accuracy of voice command recognition, the responsiveness of the command controller, and the precision of the hardware response. Adjustments were made based on test outcomes, optimizing command processing times and improving the accuracy of the hardware control mappings.

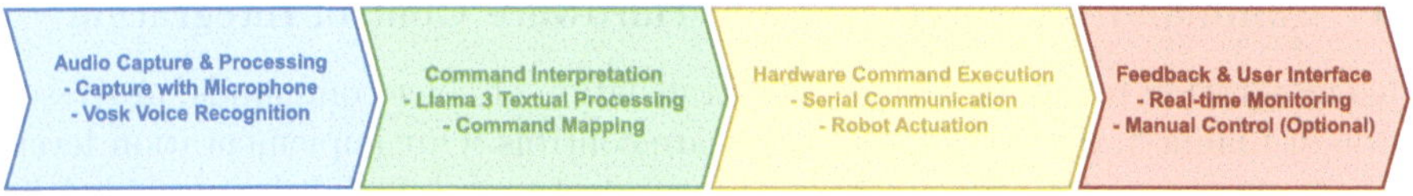

Fig. 2. Flowchart of the voice command processing system

4 Results and Discussion

The implementation and integration of the ARMOS TurtleBot with the Ollama framework and Vosk-based voice recognition have yielded promising results. Initial tests conducted in a controlled environment demonstrated that the system effectively interprets and executes voice commands with a high degree of accuracy. For example, the robot successfully navigated complex obstacle courses using only verbal instructions such as "*move forward*," "*turn left*," and "*stop*." The response times for command recognition and execution were consistently below three seconds, indicating a real-time capability suitable for dynamic environments.

To evaluate the performance of the voice recognition system, we conducted a series of tests using a dataset that included both male and female voices across a range of ages. The system processed a total of 500 commands, and the results were categorized into three groups: correctly recognized commands, incorrectly recognized commands, and unintelligible commands. Table 3 summarizes the performance:

Table 3. Performance of Voice Recognition System

Category	Number of Commands	Percentage (%)
Correctly Recognized	430	86%
Incorrectly Recognized	50	10%
Unintelligible	20	4%

The results indicate that the system achieved an 86% success rate in correctly recognizing commands, with a 10% error rate and 4% of commands deemed unintelligible. A statistical analysis using a confidence interval of 95% confirmed the reliability of these results, suggesting that the voice recognition system is effective in dynamic environments with varying levels of background noise. The data also highlight the importance of using a high-sensitivity microphone and advanced noise-canceling algorithms to enhance recognition accuracy.

Additionally, while the voice processing was performed on a dedicated PC, preliminary tests were conducted to assess the feasibility of using a Raspberry Pi 4B for this purpose. The results showed that the Raspberry Pi could handle the computational load for both the Llama 3 model and Vosk voice recognition with reasonable latency. However, during the experimental phase, the PC-based solution provided more consistent performance and flexibility for managing complex scenarios. Future iterations may explore optimizing the software to run efficiently on the Raspberry Pi, potentially increasing the system's overall portability and autonomy.

The training dataset used for this research was carefully curated to include a balanced representation of both male and female voices across different age groups. This diversity was intended to ensure the robustness of the voice recognition model in real-world applications where the system may encounter a wide range of users. The inclusion of female voices, which accounted for 50% of the training data, was particularly important given the tendency of some voice recognition systems to perform less accurately with female speakers. By ensuring an equitable distribution of voice samples, the system achieved a consistent level of accuracy across different demographics.

The integration of the Vosk voice recognition with the Llama 3 processing model proved to be highly effective. The system was able to differentiate between similar-sounding commands and execute the correct actions, which is crucial for practical applications where precision and reliability are paramount. Moreover, the use of natural language processing allowed for a more intuitive interaction between the users and the robot, significantly reducing the learning curve associated with operating such advanced robotic systems.

One of the notable outcomes of this project was the robot's ability to adapt its behavior based on real-time environmental feedback. For instance, during a test where the robot encountered an unexpected obstacle, it was able to halt and request further instructions without human intervention. This adaptability is facilitated by the seamless integration of sensory data and voice command processing, highlighting the potential of this technology in real-world applications such as assistive robotics and automated delivery systems.

However, several challenges were encountered during the implementation phase. The most significant was the occasional misinterpretation of commands in noisy environments. This issue was mitigated by enhancing the acoustic models of Vosk and adjusting the audio processing settings to better filter out background noise. Additionally, the initial setup required careful calibration of the audio input devices to optimize the system for different operational environments.

Further discussions with end-users and stakeholders during the field tests provided valuable insights into the system's operational effectiveness. The feedback was overwhelmingly positive, with particular praise for the system's ease of use and the quick adaptation of the robot to new commands. Suggestions for future improvements included the enhancement of the user interface and the expansion of the command vocabulary to cover a broader range of actions and interactions.

The research conducted provides a solid foundation for the further development of voice-controlled autonomous navigation systems. The successful deployment of these systems in ARMOS TurtleBot not only demonstrates the feasibility of such integrations but also opens up new avenues for enhancing human-robot interaction in various service industries. Future work will focus on refining the technology, expanding its application range, and exploring its potential in other areas of robotics and automation.

5 Conclusion

This study has successfully demonstrated the integration of advanced voice recognition and natural language processing technologies with autonomous robotic navigation through the development and deployment of the ARMOS TurtleBot equipped with the Ollama framework. The primary achievement of this research was the establishment of a reliable and intuitive system that allows for real-time verbal command navigation, significantly enhancing the interactive capabilities of service robots in dynamic environments.

The implementation of the Vosk API and Llama 3 model proved to be effective in processing and interpreting natural language commands. This enabled the ARMOS TurtleBot to perform complex navigational tasks with a high degree of accuracy and minimal latency. The robot's performance in controlled tests and field trials demonstrated its capability to adapt to new and changing environments swiftly, a critical feature for practical applications in industrial, service, and assistive technologies.

Moreover, the research addressed several key challenges associated with voice-controlled robotics, such as noise interference and command ambiguity. Solutions involving enhanced acoustic modeling and adaptive audio processing algorithms were developed to mitigate these issues, resulting in improved system robustness and reliability. The feedback from end-users highlighted the system's potential to transform how humans interact with robots, making the technology accessible and practical for everyday applications.

Looking forward, the integration of more sophisticated contextual understanding and predictive capabilities into the ARMOS TurtleBot's operational framework could further improve its effectiveness. Expanding the command vocabulary and refining the interaction models based on user feedback will also be crucial in ensuring that the system can meet a wider range of needs and preferences.

This project not only advances the field of robotic navigation but also contributes to the broader discourse on human-robot interaction. By bridging the gap between human command and robotic action, we have opened new possibilities for the application of autonomous systems across various sectors. Continued research and development will focus on leveraging these insights to enhance the adaptability, efficiency, and usability of robotic systems, ultimately driving forward the integration of robots into human-centric environments.

Acknowledgements. The authors gratefully acknowledge the support from Universidad Distrital Francisco José de Caldas, particularly the Facultad Tecnológica, which was crucial for this research. Special thanks are extended to the students and researchers of the ARMOS research group for their invaluable contributions. The opinions, findings, and conclusions presented in this paper are those of the authors and do not necessarily reflect the views of the Universidad Distrital. We also appreciate all collaborators for their essential roles in the realization of this project.

References

1. Alatise, M., Hancke, G.: A review on challenges of autonomous mobile robot and sensor fusion methods. IEEE Access **8**(1), 39830–39846 (2020). https://doi.org/10.1109/access.2020.2975643
2. El Asmar, B., Chelly, S., Azzi, N., Nassif, L., Asmar, J.E., Färber, M.: AWARE: a situational awareness framework for facilitating adaptive behavior of autonomous vehicles in manufacturing. In: Janowicz, C., et al. (eds.) ISWC 2020. LNCS, vol. 12507, pp. 651–666. Springer, Cham (2020). https://doi.org/10.1007/978-3-030-62466-8_40
3. Austin, M., Brown, J., Nicholson, J., Clark, J.: Dyno-kinematic leg design for high energy robotic locomotion. J. Mech. Robot. **15**(3), 1–19 (2022). https://doi.org/10.1115/1.4055998

4. Badr, A., Abdul-Hassan, A.: A review on voice-based interface for human-robot interaction. Iraqi J. Electr. Electron. Eng. 16(2), 91–102 (2020). https://doi.org/10.37917/ijeee.16.2.10
5. Beck, A., Andersen, N., Andersen, J., Ravn, O.: Mobotware: a plug-in based software framework for mobile robots. IFAC Proc. Vol. **43**(16), 127–132 (2010). https://doi.org/10.3182/20100906-3-IT-2019.00024
6. Bijjahalli, S., Sabatini, R., Gardi, A.: Advances in intelligent and autonomous navigation systems for small UAS. Prog. Aerosp. Sci. **115**(1), 100617 (2020). https://doi.org/10.1016/j.paerosci.2020.100617
7. Blass, T., Hamann, A., Lange, R., Ziegenbein, D., Brandenburg, B.: Automatic latency management for ROS 2: benefits, challenges, and open problems. In: 2021 IEEE 27th Real-Time and Embedded Technology and Applications Symposium (RTAS). IEEE (2021). https://doi.org/10.1109/rtas52030.2021.00029
8. Chung, T., Gunhee, K., Munsang, K., Chong-Won, L.: Integrated navigation system for indoor service robots in large-scale environments. In: Proceedings of IEEE International Conference on Robotics and Automation ICRA 2004, pp. 3425–3430. IEEE (2004). https://doi.org/10.1109/ROBOT.2004.1302526
9. Farouk, M.: Studying human robot interaction and its characteristics. Int. J. Comput. Inf. Manuf. (IJCIM) **2**(1), 1–16 (2022). https://doi.org/10.54489/ijcim.v2i1.73
10. Fottner, J., et al.: Autonomous Systems in Intralogistics – State of the Art and Future Research Challenges, 2 edn. Bundesvereinigung Logistik (BVL) e.V. (2021). https://doi.org/10.23773/2021_2
11. Fragapane, G., de Koster, R., Sgarbossa, F., Strandhagen, J.: Planning and control of autonomous mobile robots for intralogistics: literature review and research agenda. Eur. J. Oper. Res. **294**(2), 405–426 (2021). https://doi.org/10.1016/j.ejor.2021.01.019
12. Geihs, K.: Engineering challenges ahead for robot teamwork in dynamic environments. Appl. Sci. **10**(4), 1368 (2020). https://doi.org/10.3390/app10041368
13. Grumeza, T.-R., Lazăr, T.-A., Fortiş, A.-E.: Social robots and edge computing: integrating cloud robotics in social interaction. In: Barolli, L. (ed.) AINA 2024. Lecture Notes on Data Engineering and Communications Technologies, vol. 204, pp. 55–64. Springer, Cham (2024). https://doi.org/10.1007/978-3-031-57942-4_7
14. Koh, S., et al.: Real-time deep reinforcement learning based vehicle navigation. Appl. Soft Comput. **96**(1), 106694 (2020). https://doi.org/10.1016/j.asoc.2020.106694
15. Marge, M., et al.: Spoken language interaction with robots: recommendations for future research. Comput. Speech Lang. **71**(1), 101255 (2022). https://doi.org/10.1016/j.csl.2021.101255
16. Martínez, F., Hernández, C., Rendón, A.: Identifier of human emotions based on convolutional neural network for assistant robot. TELKOMNIKA Telecommun. Comput. Electron. Control **18**(3), 1499 (2020). https://doi.org/10.12928/telkomnika.v18i3.14777 https://doi.org/10.12928/telkomnika.v18i3.14777 https://doi.org/10.12928/telkomnika.v18i3.14777
17. Martínez, F., Montiel, H., Martínez, F.: A novel visual tracking scheme for unstructured indoor environments. Int. J. Electr. Comput. Eng. (IJECE) **13**(6), 6216 (2023). https://doi.org/10.11591/ijece.v13i6.pp6216-6227 https://doi.org/10.11591/ijece.v13i6.pp6216-6227
18. Mo, Y., Ma, S., Gong, H., Chen, Z., Zhang, J., Tao, D.: Terra: a smart and sensible digital twin framework for robust robot deployment in challenging environments.

IEEE Internet Things J. **8**(18), 14039–14050 (2021). https://doi.org/10.1109/jiot.2021.3068736
19. Mohebbi, A.: Human-robot interaction in rehabilitation and assistance: a review. Curr. Robot. Rep. **1**(3), 131–144 (2020). https://doi.org/10.1007/s43154-020-00015-4
20. Niloy, M., et al.: Critical design and control issues of indoor autonomous mobile robots: a review. IEEE Access **9**(1), 35338–35370 (2021). https://doi.org/10.1109/access.2021.3062557
21. Omeed, H., Alani, A., Rasul, I., Ashir, A., Mohammed, S.: Integrating computer vision and language model for interactive AI - robot. In: 2024 21st International Multi-Conference on Systems, Signals & Devices (SSD). IEEE (2024). https://doi.org/10.1109/ssd61670.2024.10548649
22. Park, J., Baeg, S., Baeg, M.: An intelligent navigation method for service robots in smart environments. In: Proceedings of the Control, Automation and Systems 2007 ICCAS '07, pp. 494–499. IEEE (2007). https://doi.org/10.1109/ICCAS.2007.4406960
23. Prati, E., Peruzzini, M., Pellicciari, M., Raffaeli, R.: How to include user experience in the design of human-robot interaction. Robot. Comput.-Integr. Manuf. **68**(1), 102072 (2021). https://doi.org/10.1016/j.rcim.2020.102072
24. Qian, H., Ma, X., Dai, X., Fang, F.: Socially acceptable pre-collision safety strategies for human-compliant navigation of service robots. Adv. Robot. (RSJ) **24**(13), 1813–1840 (2010). https://doi.org/10.1163/016918610X527176
25. Ragavan, S., Ganapathy, V.: A general telematics framework for autonomous service robots. In: Proceedings of the 2007 IEEE Conference on Automation Science and Engineering, pp. 675–680. IEEE (2007). https://doi.org/10.1109/COASE.2007.4341840
26. Sarker, S., Jamal, L., Ahmed, S., Irtisam, N.: Robotics and artificial intelligence in healthcare during COVID-19 pandemic: a systematic review. Robot. Autonom. Syst. **146**(1), 103902 (2021). https://doi.org/10.1016/j.robot.2021.103902
27. Sikorski, P., et al.: Deployment of NLP and LLM techniques to control mobile robots at the edge: a case study using GPT-4-turbo and llama 2. arXiv preprint arXiv:2405.17670v2 (2024)
28. Truong, X., Yoong, V., Ngo, T.: RGB-D and laser data fusion-based human detection and tracking for socially aware robot navigation framework. In: Proceedings of the 2015 IEEE International Conference on Robotics and Biomimetics (ROBIO 2015), pp. 1143–1148. IEEE (2015). https://doi.org/10.1109/ROBIO.2015.7418835
29. Udupa, S., Kamat, V., Menassa, C.: Shared autonomy in assistive mobile robots: a review. Disabil. Rehabil. Assist. Technol. **18**(6), 827–848 (2021). https://doi.org/10.1080/17483107.2021.1928778
30. Yang, L., Jiang, J., Gao, X., Wang, Q., Dou, Q., Zhang, L.: Autonomous environment-adaptive microrobot swarm navigation enabled by deep learning-based real-time distribution planning. Nat. Mach. Intell. **4**(5), 480–493 (2022). https://doi.org/10.1038/s42256-022-00482-8
31. Yasuda, Y., Martins, L., Cappabianco, F.: Autonomous visual navigation for mobile robots: a systematic literature review. ACM Comput. Surv. **53**(1), 1–34 (2020). https://doi.org/10.1145/3368961
32. Zhang, M., Xu, R., Wu, H., Pan, J., Luo, X.: Human-robot collaboration for on-site construction. Autom. Constr. **150**(1), 104812 (2023). https://doi.org/10.1016/j.autcon.2023.104812

Data Warehouse for Monitoring Patients Served by a Network of Public Health Centers

Wilson Castillo-Rojas(✉), Francisco Alcota Puga, and Victor Ángulo Olivares

Departamento de Ingeniería Informática y Cs. de la Computación, Universidad de Atacama, Av. Copayapu 485, Copiapó 1530000, Chile
wilson.castillo@uda.cl,
{francisco.alcota.18,victor.angulo.19}@alumnos.uda.cl

Abstract. This article describes the design and implementation of a data warehouse that provides indicators on patient care for a network of public health centers in a Latin American country. The aim of this work is to provide a platform to support the decision-making of strategic users responsible for managing this healthcare network. Additionally, it validates a new methodology that combines different approaches, techniques, tools, and methodologies, such as information requirements specification, relational database modeling, and a combined development process based on the Kimball and Hephaestus methodologies. This process also incorporates a validation phase of indicators and integrated, interactive visualizations for the multidimensional analysis of the obtained indicators. Regarding the results obtained with the implementation of this data warehouse, it is noteworthy that the institution in charge of managing the health centers now has a platform that allows them to analyze the average rates of appointments and patient care, as well as the costs associated with these appointments, among other indicators. This enables them to more clearly identify health centers prone to patient care collapse and take preventive measures to avoid it. From the perspective of the data warehousing process development, the team responsible for it confirms the simplicity and effectiveness of the new methodological proposal.

Keywords: business intelligence · data warehouse · data mart · data warehousing process · health centers

1 Introduction

Patient care in public health centers in Latin America exhibits significant variability depending on the country and specific region. However, there are general shared aspects, such as limited access to medical care in rural or marginalized areas due to insufficient infrastructure, human resources, and financial resources [1].

F. Suni-Lopez et al. (Eds.): CIIS 2024, ISEM 48, pp. 68–90, 2025.
https://doi.org/10.1007/978-3-031-91481-2_5

Another crucial aspect is the disparity in the quality of care, with some places offering high-quality services and others showing deficiencies due to the country's level of development, the allocation of state funding, the shortage of trained personnel, and the obsolescence of medical equipment, among other factors [2].

A recurring problem in Latin American countries is the scarcity of resources, as they face budget restrictions that limit their ability to provide quality medical care. This can manifest itself in a lack of medications, medical equipment, and trained personnel, which directly affects the quality and availability of health services [3].

Despite efforts in many countries in the region to strengthen primary health care and adopt a preventive approach to address diseases before they become serious problems, these advances are slowed by resource scarcity and a high burden of chronic diseases. The Cuban family health model stands out; however, it faces difficulties such as the segmentation of health services, a consequence of the neoliberal system. It is essential to strengthen primary health care with a family and community approach, integrating services at three levels. This involves promoting primary prevention, fostering community empowerment, allocating resources appropriately, and strengthening mental and sexual health programs more comprehensively. These actions must be based on implemented strategies and adapted to local needs to achieve a more effective and equitable health system [4].

The COVID-19 pandemic has significantly exacerbated challenges already present in Latin America's public health systems, such as shortages of medical personnel and personal protective equipment, in addition to disrupting health services unrelated to the virus. Therefore, it is crucial to monitor patient care in public health facilities to effectively manage the limited resources available in these facilities [4,5].

In this context, the main objective of the article is to develop a data warehouse (DW), to monitor patient care in various public health centers in a Latin American country. Additionally, it seeks to validate the effectiveness of a new methodology for the data warehousing process, detailed in [6], as a secondary objective.

A fundamental part of business intelligence technology, DWs have evolved with the emergence of new data types and storage methods. A DW integrates data from multiple sources into a single centralized repository, supporting data analysis and making it easier for organizations to manage large volumes of historical data. The concept of DW was introduced by Inmon and Hackathorn in the 1990s, defining it as a topic-oriented collection of data with the purpose of assisting management decision-making in the various business processes of an organization [7].

Several authors support Kimball's definition about a data mart, highlighting its difference from a DW, both in terms of data storage and analysis. According to these authors, a data mart is a smaller and specialized version of a DW, focused on a specific department or business area. Although its goal is to facilitate decision-making and data analysis, its scope is more limited, and its structure can be simpler than that of a DW [8,9].

In the design of a DW or data mart, three multidimensional models are distinguished: star, snowflake, and constellation. The star model organizes data in a centralized structure, with a fact table and multiple dimension tables, making it efficient for analysis. On the other hand, the snowflake model normalizes dimension tables into a hierarchical structure to reduce redundancies, although it is more complex. Finally, the constellation model combines star and snowflake characteristics, offering flexibility and being suitable for complex DWs with various data sources [7–9].

To create the DW, various pieces of information are collected about multiple health centers, their patients, and the consultations conducted in said establishments. These data are obtained from records of various neighborhoods or districts located in different cities of the country. It is crucial to integrate this dispersed information into various operational databases to achieve centralized management of patient data. This will allow for obtaining a global view of health services and calculating key performance indicators (KPIs) for proper comparative analysis.

The objective of the presented DW is to enhance the available information, providing a comprehensive vision of the various aspects that comprise the patient care process in the different public health centers of the country. This aims to increase efficiency in care and facilitate strategic decision-making to promote sustained development within a process of continuous improvement. For example, the goal is to determine the average rate of patient care in each health center in different cities, as well as the average costs associated with these consultations, among other KPIs.

The new methodology used for the development of this DW combines the approaches of Kimball and Hephaestus [6]. On one hand, Kimball's methodology follows a series of steps covering the selection of the business process, architecture design, identification of dimensions and facts, construction of dimensional models, and gradual implementation of the DW. This approach prioritizes incremental delivery and collaboration between users and developers to ensure alignment with business requirements and continuous delivery of value [10]. On the other hand, the Hefesto methodology establishes guidelines aimed at guaranteeing the quality of the DW development process and improving project efficiency. It includes defining the scope, identifying requirements, designing the data model, implementing the Extract, Transform, Load (ETL) process, and creating reports for analysis [11].

For the implementation of the DW, a multidimensional relational database model is used, integrating data from various sources such as the patient appointment registration system and records of care costs by specialty. This data undergoes an ETL process using the Pentaho Data Integration tool, which enables the design, documentation, and generation of workflows for this process, ensuring the consistency and quality of the integrated data. The DW has been designed to generate analysis reports on various aspects related to the management of patient appointments in the different public health centers of the analyzed health network. To view these analysis reports and their corresponding graphs, the data analytics tool Tableau is utilized.

As a result of this work, the DW provides the organizations responsible for managing public health centers with a more comprehensive understanding of their capacity to care for patients within the public health network, facilitating decision-making and the identification of areas for improvement. This enables the monitoring of the effectiveness of appointment programs, as well as patient care and support services. The implementation of a DW for this purpose is a valuable initiative that contributes to the continuous improvement of a country's health system. The results of the KPIs are presented through graphs and tables, allowing for the examination of appointment management and patient care by specialty and type of emergency, among other aspects, over different periods, grouped by health centers, specialties, and communities. From the point of view of the data warehousing process, the team in charge of development confirms the usefulness, simplicity, and effectiveness of the new methodology employed.

2 Related Works

After the pandemic, public services, especially in Latin American countries, have had to adapt to maintain efficiency in patient care. The need to incorporate information technologies into the healthcare sector has resurfaced, highlighting the role of DW systems in measuring and improving processes. Before developing these systems, it is essential to define KPIs and utilize business intelligence tools for analyzing institutional data in all healthcare areas.

There is a diversity of works that support this trend and enable us to observe the different objectives, methodologies, techniques, and computational tools used. In this section, an analysis of some related works found in different scientific databases is presented.

Jiménez's study [12] examines quality and efficiency in medical care, considering indicators and challenges for their evaluation. The definition of quality in health is complex and diverse, closely linked to efficiency according to health administrators. Measuring these aspects is challenging due to their abstract nature and the different interests involved. Health centers play a key role in treating serious and expensive patients, underscoring the importance of continually evaluating quality and efficiency for their social and economic impacts.

Barriga-Chambi et al. [13] evaluate the satisfaction of health personnel and patients in a Peruvian hospital, as well as the implementation of telehealth, especially relevant due to the COVID-19 pandemic. Despite its advantages, such as greater access to healthcare, it faces challenges such as regulatory gaps and the digital divide. They highlight the importance of patient and health personnel satisfaction, influenced by the quality of service and the doctor-patient relationship, although satisfaction among health personnel has been less studied. In Latin America and the Caribbean, the evaluation of satisfaction in telehealth is limited, highlighting the need for more research to improve the regulation and quality of service.

Vargas and Hernández, in their article [14], highlight the relevance of information systems in hospital management, emphasizing their usefulness for making decisions aligned with social objectives. Their analysis, based on a theoretical-bibliographic review and the legal framework, concludes the need to strengthen these systems with trained personnel for their implementation. The importance of ensuring the reliability, relevance, and timeliness of the results generated is emphasized, as well as their effective use in hospital management, prioritizing the optimization of resource utilization from a social perspective rather than economic profitability.

Valeri [15] highlights the crucial importance of information systems in public health management, which convert data into useful information for decision-making. These systems are essential for the effective functioning of health services and contribute to the management of service delivery. It is necessary that the information generated is aligned with organizational objectives and contributes to their achievement, ensuring accurate data to improve the efficiency and quality of processes and to control the allocation of resources. All these aspects must be reflected in the planned budget of the state health system, meeting premises of equity, effectiveness, efficiency, and user satisfaction.

The study in [16] seeks to advance second-generation marketing by evaluating the perceptions of officials and patients on specific variables in health centers. It highlights the importance of focusing on improvements from the patient's perspective to meet their expectations and needs, proposing a personalized application of individual priorities in health system reforms. Although some measurement challenges have been overcome, reservations remain about a partial view focused solely on the patient.

Castañeiro et al. in [17] present the XAVIA HIS system, a comprehensive solution for medical management in hospitals, including a specific module for outpatient consultation that manages various specialties. Their functionalities and medical specialties stand out, strengthening the knowledge base for clinical and administrative decisions, thereby improving the quality of patient care.

Bocquet et al. in [18] highlight the impact of big data and artificial intelligence in oncology, focusing on the need to develop effective DWs for cancer research. Although they promote the use of DWs to integrate data and address challenges, they point out technical, organizational, and legal concerns, focusing on public health risks and legal aspects for real-world evidence studies.

Granda et al. in [19] present a DW model developed in the zonal health coordination of Ecuador, highlighting the importance of data analysis in public health decision-making. Their objective is to generate organized information for efficient analysis, concluding that a well-implemented DW can enhance the capacity of health institutions to use data in decision-making.

A concise comparative analysis of the most prevalent methodologies for DW development indicates that the Kimball methodology, also known as dimensional modeling, is among the most widely utilized. This approach is founded upon a design that prioritizes comprehensibility and high performance, emphasizing the creation of data cubes and fact and dimension tables. In comparison to other methodologies, Kimball is more straightforward and expeditious to implement, although it may not be optimal for large-scale or highly intricate systems [10].

In contrast, the Hefesto methodology, which is based on a reverse engineering approach, is primarily concerned with the extraction of information from disparate sources, followed by its transformation and loading into the DW. Hefesto is distinguished by its flexibility in data integration, which renders it particularly well-suited to legacy systems [11].

The Inmon methodology, otherwise known as enterprise data modeling, is characterized by the generation of a normalized data model that is incorporated into a corporate DW. Although more complex and costlier to implement than the Kimball methodology, it offers superior data integrity and consistency [20].

Finally, the Data Vault method represents a hybrid approach that incorporates aspects of both Kimball and Inmon models, prioritizing flexibility, scalability, and adaptability. This approach permits the management of substantial quantities of data and is readily adaptable to evolving business environments, facilitating expeditious modifications to the data model [21].

The analyzed works describe the importance of implementing computer systems and DW solutions in public health environments to enhance patient care efficiency. By having access to patient records, appointments, and conducting data analysis, these systems can contribute to better decision-making regarding resource management in health centers.

The studies reviewed emphasize the significance of introducing IT and DW solutions in public health settings to optimize patient care. Access to patient and appointment records, along with data analysis, facilitates decision-making related to resource management in health centers.

From a technological perspective, various initiatives for the implementation of DW systems are observed in several countries, highlighting the use of multidimensional models and the Kimball methodology for these solutions in the healthcare field. Similarly, the importance of defining KPIs and determining the relevant variables for data analysis is emphasized. In all these projects, the ETL process phase is essential for integrating data from multiple sources, calculating KPIs, and loading them into the DW.

Some studies detail the implementation of analytical tools and control panels, which allow strategic users to visualize information about patient care in health centers clearly and concisely. Together, these works illustrate how the implementation of DW solutions can support decision-making in health centers, thus contributing to improving their efficiency and the quality of the services offered.

After examining the related works, the following conclusions can be drawn:

- The incorporation of DW solutions in this sector can have a significant impact on knowledge management and strategic decision-making to improve care in public health centers.
- Methodologies based on Kimball and Hefesto are widely used in the implementation of these solutions individually, representing a solid option to ensure success in the development of data warehousing processes.

- The introduction of DW systems in the healthcare field expands the capacity for analysis and visualization of information, potentially contributing to improved efficiency and effectiveness of processes in health centers, including appointment and patient care management.
- Some studies highlight the importance of ethics in the use of data and the need to protect patient privacy.

3 Materials and Method

3.1 Development Methodology

The methodology used in this work is based on the approach detailed in [6], which integrates various approaches, techniques, and methodologies. It includes the specification of information requirements-a common practice in software engineering. It also integrates the expanded ETL process, enriched with an indicator validation phase (ETL+V), as well as integrated and interactive visualizations for multidimensional analysis of indicators, based on the dashboard concept.

This methodology is built on the experience accumulated in the development of data warehousing projects, which focus on generating indicators or KPIs that evaluate the performance and effectiveness of business processes in various organizations.

The creation of this methodology is based on the conventional development model for DW by Kimball & Ross [7], which is merged with the Hefesto methodology by Bernabeu [8], extracting the common phases and combining them. Kimball's approach is based on the dimensional business life cycle, covering stages such as the definition of business requirements, dimensional modeling, physical design, and implementation of the ETL subsystem. On the other hand, the Hefesto methodology consists of four stages: requirements analysis, operational systems analysis (online transaction processing), DW logical modeling, and data integration.

In the new methodology, the first stages of both methodologies are unified into a single initial stage called Analysis, which also incorporates the second stage of Hefesto, online transactional processing analysis. Then, Kimball's dimensional modeling and physical design stages are merged with the third stage of Hefesto, logical model of the DW, giving rise to the second stage called Multidimensional Model Design.

In the third stage, called the ETL+V Process, the design and implementation stage ETL subsystem of the Kimball is combined with the Hefesto data integration stage. The fourth stage, Analytical Processing, results from the fusion between Kimball's proposals in his life cycle of a DW and those of several authors who consider online analytical processing (OLAP) technology as an option for tool exploitation of the DW, among others, and not as part of its development method.

Each of these stages has its inputs and outputs defined in a bidirectional and interactive process that allows for returning to the previous stage or advancing to the next without losing the meaning of the DW process. The relationship and participation of users are also established, both at the entrance to the process and at the exit. Key elements stand out, such as the ad-hoc template for specifying the KPIs, the temporary repository as a storage area (staging), the SQL code for the validation of the KPIs, and the graphical visualizations integrated through control panels or dashboards [6].

3.2 Healthcare Context

Analysis of related works has highlighted the urgency of improving the efficiency of primary healthcare services for patients. This not only requires the allocation of financial resources, the recruitment of additional medical personnel, the improvement of infrastructure, and the acquisition of medical equipment; but also, the adoption of solutions that make use of modern information technologies. Data warehouse systems have stood out as one of the most widely developed technologies in various countries in the region. This is because DWs allow processes to be measured by providing a set of KPIs, which, in turn, offer a comprehensive view of the management of patient care in the health network in different communities of a country.

Among the main advantages of having a DW system are that it allows health centers to track their patients' appointments and care, and it provides information on the number of patients treated in different cities, communities, and regions, as well as by type of care or health center.

3.3 KPI Specification

Through a strategic process of information requirement induction, the need to implement a DW that stores KPIs to measure the average rate of appointments and patient care in health centers is identified. This will allow for detailed analysis and detection of critical situations requiring timely intervention. Proactive management of solutions based on these indicators is essential to improve the functioning of health centers and prevent potential collapses. In this context, examining the relationships between patients and healthcare providers is essential.

For this work, a database covering a 10-year period, from 2002 to 2022, is available. In the analysis phase, the information requirements elicitation technique is employed, commonly used in the software engineering process. This technique enables the identification and specification of the following requirements:

- Number of appointments made by patients.
- Total cost of appointments by patient.
- Analyze the number of patients according to different segmentation criteria.

To meet this last requirement, the number of patients is analyzed according to various criteria, which are detailed below. The structure of the template used for the KPI specifications is developed taking as reference the templates and linguistic patterns described in [22]. Tables 1, 2, and 3 present the specifications of the identified KPIs. These specifications include not only their description and objective, but also their periodicity, the dimensions involved, and the corresponding calculation formulas.

Table 1. Number of appointments made by patients

Indicator 1	Number of appointments made by patients
Definition/Objective	Measure the number of appointments made by patients in health centers.
Periodicity	Daily, monthly, and yearly.
Source	Operational database of the healthcare network.
Target	– Health centers – Entity in charge

Dimensions	`'personas'`
	`'centros_mutualistas'`
	`'tipo_consulta'`
	`'tiempo'`

Formula	Total number of citations by user (#TCU): $\#TCU = C_i^n$ occurrences of user$_i$ in appointments
Observations	Where C represents an accountant.

3.4 Operational Systems Analysis

Once the KPIs have been specified, an analysis of the relational model of the operational database available in the public health agency is conducted. Subsequently, the respective correspondences between the tables and the data available in this database with the KPIs are established, selecting the appropriate attributes from the operational database to calculate these key indicators.

Table 4 details each of the tables and the attributes contained in the operational database. Primary keys and foreign keys are indicated by [PK] and [FK], respectively.

It is important to highlight that the prices by consultation have been obtained from an external and public data source, specifically from the national health fund, since this attribute is not registered in the operational database. This data is then added to the 'tipo_consulta' table to allow the calculation of the KPI related to appointment costs by patient.

4 Design and Implementation

From the correspondence analysis between the attributes present in the various tables of the operational database and those required to calculate the speci-

Table 2. Total cost of appointments by patient

Indicator 2	Total cost of appointments by patient
Definition/Objective	Measure the total cost of total patient appointments at health centers.
Periodicity	Daily, monthly, and yearly.
Source	Operational database of the healthcare network Table of care costs by specialty
Target	– Health centers – Entity in charge
Dimensions	'personas' 'centros_mutualistas' 'tipo_consulta' 'tiempo'
Formula	Total cost of appointments by user (CTCU): $CTCU = \sum_i^n$ user$_i$ appointment cost
Observations	Represents the sum of $user_i$ costs. n = user's total appointments.

Table 3. Number of patients based on different analysis criteria

Indicator 1	Number of patients based on different analysis criteria
Definition/Objective	Measure the number of patients according to different analysis criteria in health centers.
Periodicity	Daily, monthly, and yearly.
Source	Operational database of the healthcare network
Target	– Health centers – Entity in charge
Dimensions	'personas' 'centros_mutualistas' 'tipo_consulta' 'tiempo'
Formula	Number of patients by criteria (#PP): $\#PP = C_i^n$ occurrences of user$_i$ by criteria
Observations	It represents the number of patients for each established criterion, for example neighborhoods/districts, city, gender, specialty, date.

fied KPIs, the necessary attributes for this purpose are identified. Likewise, the required attributes are selected for the respective dimensions that allow the indicators to be described.

Table 4. Number of patients based on different analysis criteria

Tables	Columns	Description
'centros_mutualistas'	id_centro_mutualista [PK]	It contains information on names and cities where each health center is located.
	centro	
	id_mutualista [PK]	
	departamento	
'consultas'	id_persona [FK]	It contains only foreign keys and two primaries. It acts as a purely relational table, that is, as a link.
	id_centro_mutualista [FK]	
	id_tipo_consulta [FK]	
	id_diagnostico [PK]	
	fecha_consulta [PK]	
'his_person'	id_persona [FK]	Contains information about where and when each patient enters and leaves a health center.
	id_centro_mutualista [FK]	
	id_tipo_consulta [FK]	
	id_diagnostico [PK]	
	fecha_consulta [PK]	
'personas'	id_persona [FK]	Contains complete patient information.
	nombre	
	apellido	
	fecha_nacimiento	
	sexo	
	ciudad_residencia	
	ocupacion	
'tipo_consulta'	id_tipo_consulta [PK]	Contains information about the different types of queries made, along with associated information.
	modo_consulta	
	tipo_socio	
	origen_consulta	
	consulta	

4.1 Specification of Dimension Tables

For the multidimensional modeling of the DW, a star schema is chosen, commonly known as a data mart due to its characteristics, which include a set of dimensions linked to a single central fact table. Below are the descriptions of the four dimensions defined for the multidimensional model, which are necessary to describe the specified KPIs.

- 'personas': It is necessary to know the number of appointments by user, so this information is of great importance. In relation to this dimension, only the identifier and sex are used, since the other data are not considered relevant for the analysis.
- 'centros_mutualistas': This dimension refers to the health centers where people carry out their consultations. Additionally, an associated dimension called 'centro_general' is added, which includes the name of the general center to which the health center in question belongs.
- 'tipo_consulta': Corresponds to the classification of medical consultations carried out.
- 'tiempo': This dimension is fundamental for the DW, as it provides temporality to the data. In this analysis, a granularity covering day, month, and year is used, considered the most appropriate for this work.

4.2 Fact Table Specification

The fact table is defined by incorporating the selected attributes necessary for the calculation of each KPI, along with their respective formula. Table 5 presents the components of the fact table named "citas", which includes the following attributes and keys:

Table 5. Number of patients based on different analysis criteria

Columns	Description
id_personas	[FK] with the 'personas' dimension.
id_centros_mutualistas	[FK] with the 'centros_mutualistas' dimension.
id_tipo_consulta	[FK] with the 'tipo_consulta' dimension.
id_fecha	[FK] with the 'tiempo' dimension.
costo	Measure that contains the total cost of appointments by person.
cantidad_citas	Measurement indicating the number of appointments.

4.3 Multidimensional Model

Once the components of the multidimensional model, including both the facts and dimensions, have been established, the conceptual model is designed using a CASE tool. This design process entails connecting the fact table to the corresponding dimensions and, for the 'tiempo' dimension, to its respective linear hierarchy level. Figure 1 depicts the resulting multidimensional model based on the data collected during the previous analysis stage.

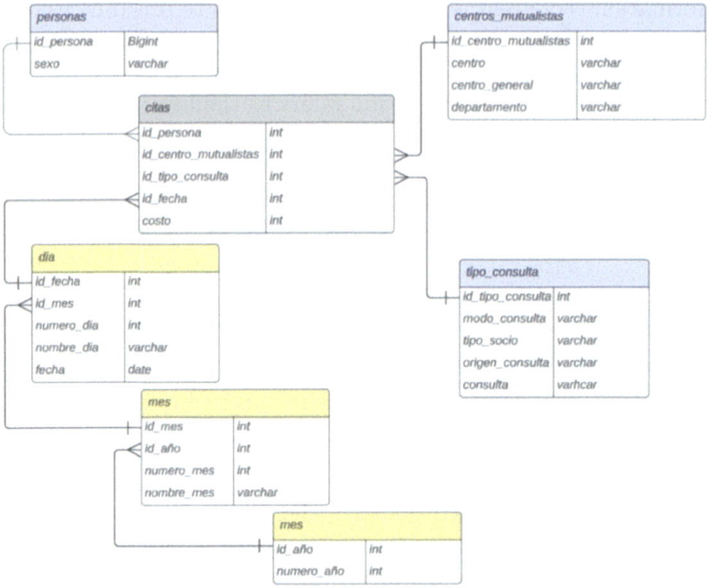

Fig. 1. Multidimensional model of the DW.

4.4 ETL Process Description

To implement the DW, the Pentaho Data Integration ETL tool is used. This tool models the ETL process through various workflow diagrams, which consist of a set of objects called "jobs" that group actions known as "transformations". These transformations are utilized to configure connections to the source data sources, extract data, perform KPI modifications or calculations, and establish connections to the target data sources for storage. It is important to note that SQL statements or Java scripts can be integrated into these transformations. In this stage, the main diagrams of the ETL process developed for this work are presented.

Before populating the dimension and fact tables, the relational physical model of the DW is generated. Additionally, selected data is extracted and migrated from the provided operational database on a server. Populating the multidimensional DW database begins with dimension tables and ends with fact tables.

Dimension Tables. For the 'tipo_consulta' dimension, a transformation is carried out to incorporate the prices associated with each type of consultation by specialty, which are extracted from an external data source. Figure 2 illustrates the flow of this transformation, in which the necessary attributes are selected, sorted, prices are assigned to the queries, and finally, they are loaded into the corresponding table.

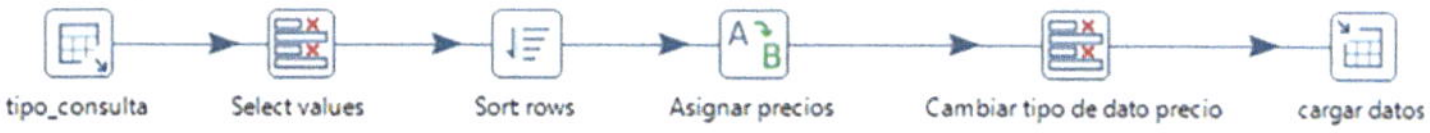

Fig. 2. Generation and loading of data in 'tipo_consulta' dimension

For the dimensions 'tipo_consulta' and 'personas', the necessary attributes are chosen from the operational database, sorted, and finally, loaded into the corresponding table of each dimension. The workflows for both dimensions are similar and quite simple, so only the workflow for the person dimension is shown in Fig. 3.

Fig. 3. Generation and loading of data in 'personas' dimension

For the 'centros_mutualistas' dimension, several transformations are carried out. One of them consists of extracting the address, which is then concatenated with the name attribute of the health center, while also combining it with the street attribute found in another separate attribute. Then, the necessary attributes are selected, some formatting is adjusted to uppercase, and duplicate entries are removed, before finally loading the data into the dimension. This process is illustrated in Fig. 4.

The 'tiempo' dimension is the one that presents the greatest complexity for its generation, as it is a hierarchical dimension that allows analysis of KPIs at different levels of granularity, with the day being the most detailed level, followed by the month, and finally the year as a highest level of granularity. This dimension must be created from the dates recorded in the operational database for each consultation made by patients during the analyzed period.

To begin, the year granularity of the 'tiempo' dimension is generated, as it is the simplest of all. It does not require foreign keys, and it is only necessary

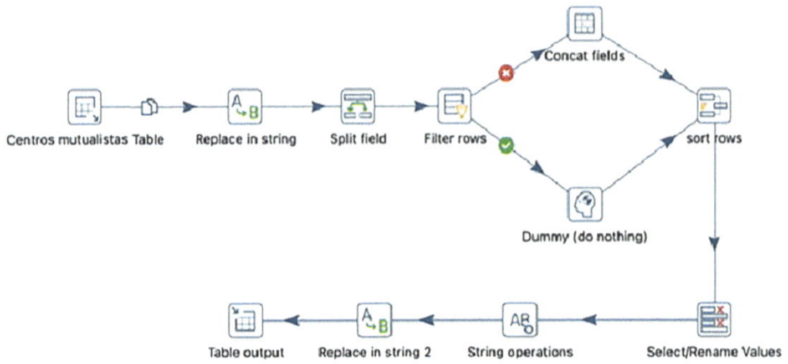

Fig. 4. Generation and loading of data in 'centros_mutualistas' dimension

to extract the years from the dates of the queries and assign them an identifier (id_año). Figure 5 presents the workflow for this task, illustrating this transformation.

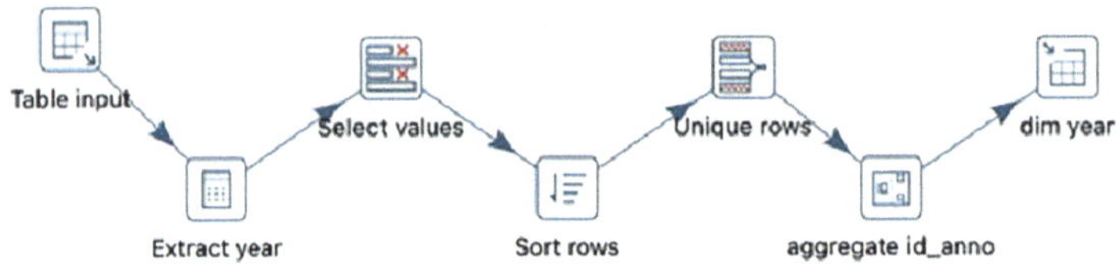

Fig. 5. Generation and loading of data in 'tiempo' dimension at year granularity

We proceed with the generation of the month granularity, which involves a little more work. The first part of the transformation is similar to the previous one, but in this case, the month is extracted instead of the year. Then, unique month values are selected and assigned their respective names. Subsequently, each month is associated with all the years present in the year dimension using conditionals that evaluate the value of 'id_año'. Next, an incremental identifier is assigned to each record in the month dimension (id_mes), and finally, the data is selected to be loaded into this dimension. The complete workflow for this month dimension can be seen in Fig. 6.

For the last granularity, day, of the 'tiempo' dimension, we begin by extracting the dates from the query table to obtain only the corresponding days. Then, they are extracted from the month granularity table, and these tables are joined to create all the existing dates in the operational database. The years are then mapped using the 'id_año' to concatenate each extracted day, month, and year value into a single date. The days of each date are associated with their respective week names. Subsequently, the necessary attributes are selected, and the data type is modified to date type. Finally, the records are loaded into the day granularity table of the time dimension.

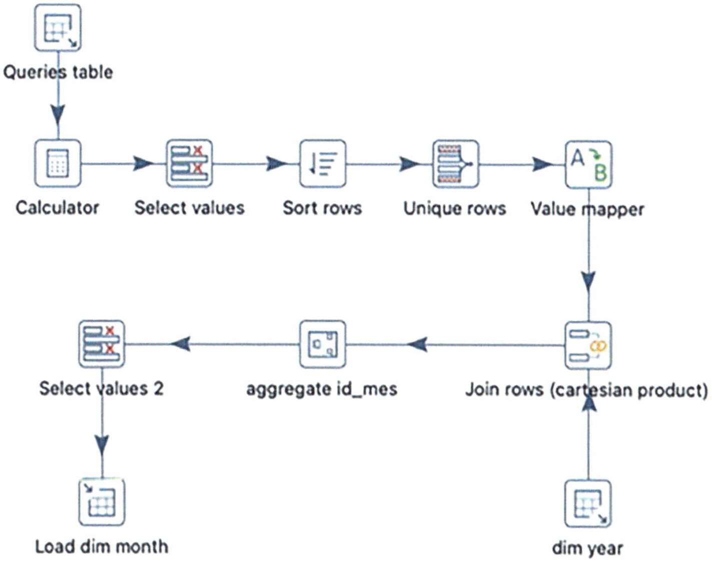

Fig. 6. Generation and loading of data in 'tiempo' dimension at year granularity

Figure 7 presents the complete workflow to generate and load the data of the day granularity of the 'tiempo' dimension, where its greater complexity can be observed compared to the other granularities of this dimension.

Fact Table. To populate the fact table called 'citas' and calculate both the first and third KPIs, the consultation data needs to be grouped by patient. Additionally, to calculate the second KPI, which involves obtaining the total cost of the appointments made by each patient, it is necessary to access the prices of each consultation according to its specialty type. To achieve this, the required data is first extracted and selected from the query table of the operational database. Then, it is related to the 'tipo_consulta' table of the DW multidimensional model, which contains the prices of each consultation type by specialty. The keys 'id_tipo_consulta' are used to relate both tables and add the price associated with each query.

Once the price is obtained, the identifier of the date associated with the query is added. This is done simultaneously with obtaining and concatenating the dates and identifiers of the 'mes' and 'año' dimensions. These steps allow for the association and addition of the necessary attributes from the 'tiempo' dimension, which include the year, month, and day identifier. Figure 8 shows the complete workflow that allows for generating KPIs and recording them in the fact table.

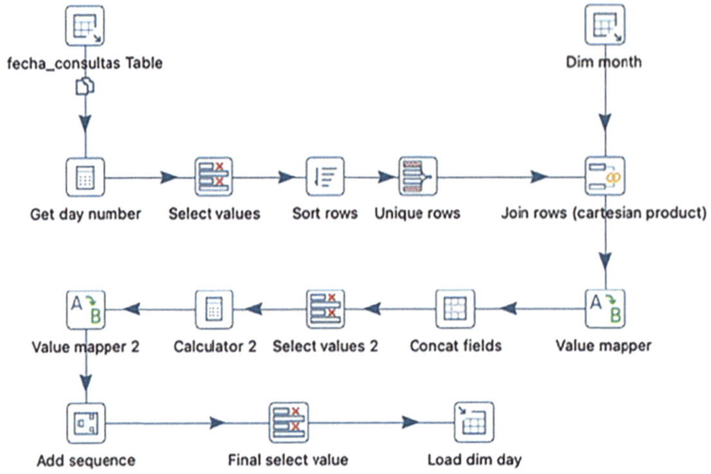

Fig. 7. Generation and loading of data in 'tiempo' dimension at day granularity

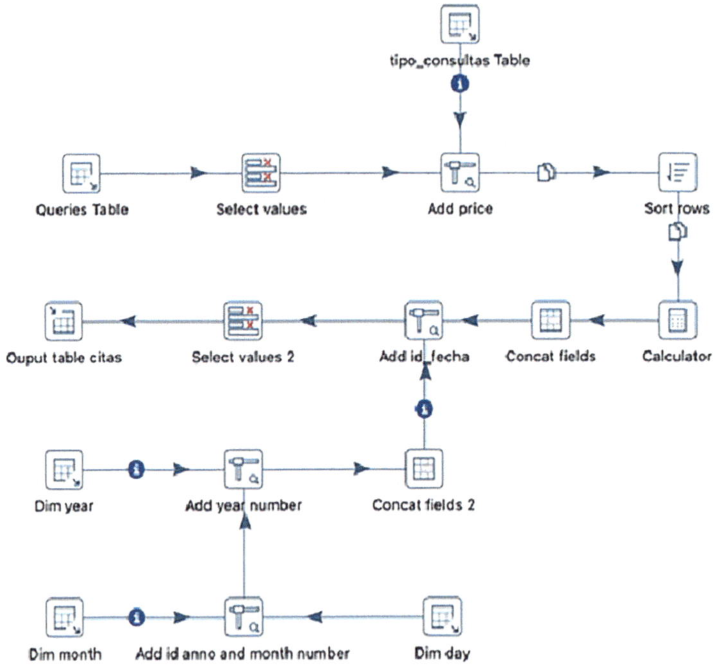

Fig. 8. Generation and loading of data into fact table 'citas'

5 Analysis and Visualization of Results

For the visualization and multidimensional analysis of the indicators, the OLAP Tableau tool in its 2023 version is used. Individual visualizations are designed for each KPI, using appropriate graphs, which are then integrated into a dashboard. This dashboard allows strategic users to access the specified indicators in an integrated manner.

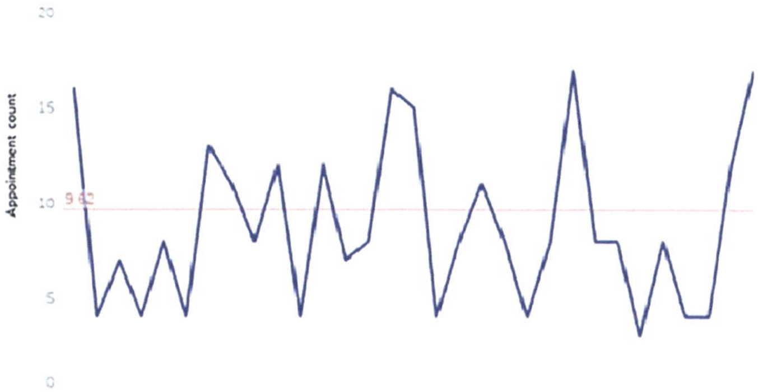

Fig. 9. Number of appointments by user in 2020

Figure 9 shows an example of a graph representing the number of appointments made by patients in a specific year, in this case 2020. It should be noted that it is possible to modify the period using the sliding filter defined for this purpose.

The graph includes the average number of appointments by user, which for the year 2020 is 9.62. It can be seen that during the first months of the year, there is a notable decrease in the number of patient appointments, coinciding with the country's summer period. However, in the following months, the number of appointments and patient care returns to a regular pattern.

Following the recommendation of the methodology used, which proposes an extended ETL process with a validation stage (+V), the corresponding validation is carried out for each KPI using SQL instructions. Below is the SQL query that validates this KPI:

```
SELECT AVG(cantidad_citas) AS promedio_citas
FROM (
SELECT id_persona, COUNT(*) AS cantidad_citas
FROM citas
JOIN dia ON citas.id_fecha = dia.id_fecha
JOIN mes ON dia.id_mes = mes.id_mes
JOIN anno ON mes.id_anno = anno.id_anno
WHERE anno.id_anno = 2020
```

```
GROUP BY id_persona
) subquery;
```

With this SQL query, it is confirmed that the average number of appointments per user for this year is approximately 9.619908, a value close to that obtained through the ETL process, which correctly validates this KPI.

For the second KPI, which calculates the total cost of appointments per user, similar steps are followed to the previous indicator. As shown in Fig. 10, the average total cost of appointments per user in 2022 is approximately \$2,433 (local currency), which is consistent with validation using SQL.

In the graph, it can be seen that there are some months in which the total cost of appointments increases considerably due to a greater number of appointments and patient care.

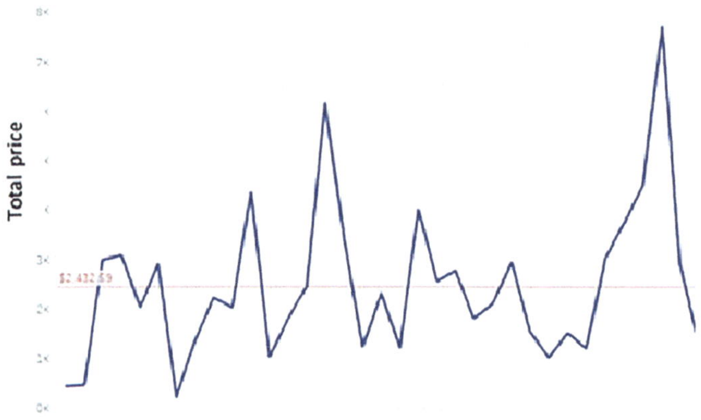

Fig. 10. Average total cost of appointments by user in 2022

The following SQL query validates this KPI and returns a similar result:

```
SELECT AVG(costo_total_por_usuario) AS promedio_costo_total_por_usuario
FROM (
SELECT id_persona, SUM(precio_total) AS costo_total_por_usuario
FROM citas
JOIN dia ON citas.id_fecha = dia.id_fecha
JOIN mes ON dia.id_mes = mes.id_mes
JOIN anno ON mes.id_anno = anno.id_anno
WHERE anno.id_anno = 2022
GROUP BY id_persona
) subquery;
```

For the third KPI, the number of appointments by year, according to the gender of each patient, is shown in Fig. 11.

To calculate this KPI, the total appointments made are divided by year and by gender of the patients, highlighting the female gender in red and the male gender in blue. As can be seen, women lead in the number of appointments throughout the period from 2002 to 2014.

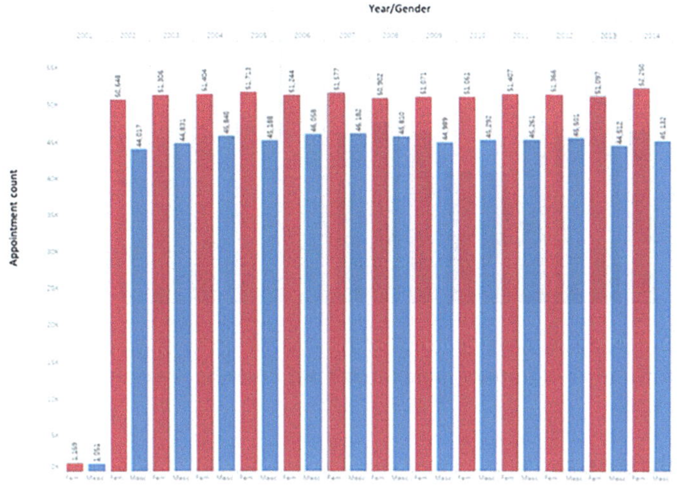

Fig. 11. Number of appointments per year by gender

Other variations of the third KPI can be seen in Fig. 12, which display the numbers of appointments and total costs divided by specialty, made during the entire analysis period. Additionally, these figures are also divided according to the type of affiliation, whether with a 'FONASA' bonus or another pension health institution labeled 'NO-FONASA'.

It is observed that the largest number of appointments corresponds to the type of home consultations, both for patients with a 'FONASA' bonus and for those 'NO-FONASA', which is related to the pandemic period between 2019 and 2020.

A brief survey was conducted among the development team to ascertain their views on the DW development methodology described in this work. The results demonstrated that the majority (more than 90%) of the team expressed high levels of satisfaction with the methodology, citing its effectiveness in streamlining the data storage process, enhancing simplicity and efficiency, and utilizing appropriate tools at each stage. In particular, the team highlighted the value of the new methodology in facilitating the design and implementation of visualizations.

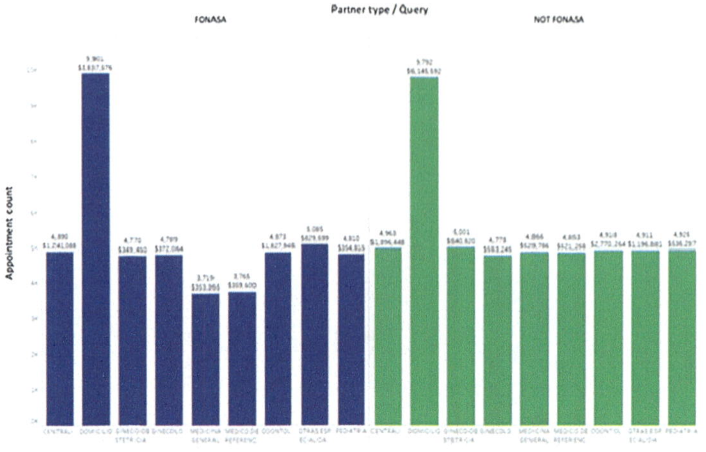

Fig. 12. Number of appointments and total costs by specialty and member

6 Conclusions

The results of this work have exceeded the expectations of strategic users and middle managers who directly use this tool. The institution responsible for managing the various public health centers now has a tool that allows it to effectively track appointments and patient care.

A customized solution has been developed to analyze the average rate of patients being attended in the different health centers, as well as the costs associated with consultations for each specialty. Additionally, appointments and patient care can be analyzed according to various dimensions, such as gender, age, and specialty.

Certainly, the design and implementation of graphical visualizations in the Tableau tool have facilitated strategic users' understanding of the defined KPIs, enabling them to interpret them easily and make confident decisions to improve care in healthcare centers and enhance patients' well-being.

From a technical standpoint, the data warehousing process conducted through the application of the new methodology, which merges stages of the Kimball and Hefesto methodologies, simplified the analysis and design work of the DW.

Furthermore, the integration of various tools such as Pentaho Data Integration for data extraction and transformation, Tableau as an OLAP tool for visualization, and PostgreSQL as a database management system, has proven to be an effective approach to address this type of project within a short timeframe.

The correspondence diagrams between the operational database and the multidimensional model of the DW recommended by the Hefesto methodology allow for the simple selection of strictly necessary data to describe and calculate the KPIs.

The importance of designing and implementing workflow diagrams for the ETL process was also evident through the use of tools such as Pentaho Data Integration. These tools enable the abstraction of the necessary programming and configuration tasks for data extraction from the operational database, its integration, the calculation of KPIs, and the loading of the multidimensional DW model.

In conclusion, this work enabled the development team to confirm the validity of the methodology and technologies recommended in [6] for successfully conducting a data warehousing process of medium complexity within a reasonable timeframe. This is in comparison to the classic methodologies typically used in such projects.

References

1. Dieleman, J.L., Sadat, N., Chang, A.Y., Fullman, N.: Trends in future health financing and coverage: future health spending and universal health coverage in 188 countries, 2016–40. Lancet **391**(10132), 1783–1798 (2018). https://doi.org/10.1016/S0140-6736(18)30697-4
2. Báscolo, E., Houghton, N., Del Riego, A.: Lógicas de transformación de los sistemas de salud en américa latina y resultados en acceso y cobertura de salud. Rev. Panam. Salud. Publica **42**, e126 (2018). https://doi.org/10.26633/RPSP.2018.126
3. Titelman, D., Cetrángolo, O., Acosta, O.L.: Universal health coverage in Latin American countries: how to improve solidarity-based schemes. Lancet **385**(9975), 1359–1363 (2015). https://doi.org/10.1016/S0140-6736(14)61780-3
4. Rojas-Torres, I.L., de Jesús Gil Herrera, R.: Main challenges of primary health care in five Latin American countries. J. Health Manag., 09720634231196950 (2023). https://doi.org/10.1177/09720634231196950
5. Ruano, A.L., Rodríguez, D., Rossi, P.G., Maceira, D.: Understanding inequities in health and health systems in Latin America and the Caribbean: a thematic series. Int. J. Equity Health **20**(1), 94 (2021). https://doi.org/10.1186/s12939-021-01426-1
6. Castillo-Rojas, W., Medina Quispe, F., Molina, F.F.: Una metodologÃa para procesos data warehousing basada en la experiencia. RISTI, 83–103 (2018). https://doi.org/10.17013/risti.26.83-103
7. Inmon, W.H., Hackathorn, R.D.: Using the Data Warehouse. Wiley-QED Publishing, USA (1994)
8. Bojičić, I., Marjanović, Z., Turajlić, N., Petrović, M., Vučković, M., Jovanović, V.: A comparative analysis of data warehouse data models. In: 2016 6th International Conference on Computers Communications and Control (ICCCC), pp. 151–159 (2016). https://doi.org/10.1109/ICCCC.2016.7496754
9. Golfarelli, M., Rizzi, S.: Data Warehouse Design: Modern Principles and Methodologies. McGraw Hill Professional, USA (2009)
10. Kimball, R., Ross, M.: The Data Warehouse Toolkit: The Definitive Guide to Dimensional Modeling. Wiley, New York (2013)
11. Bernabeu, R.D., García Mattío, M.: Hefesto - data warehousing methodology. https://troyanx.com/Hefesto/. Accessed 18 Dec 2024
12. Jiménez Paneque, R.E.: Indicadores de calidad y eficiencia de los servicios hospitalarios: Una mirada actual. Revista Cubana de Salud Pública **30**(1), 17–36 (2004). http://scielo.sld.cu/scielo.php?pid=S0864-34662004000100004&script=sci_arttext

13. Barriga-Chambi, F., et al.: Satisfaction of healthcare workers and patients regarding telehealth service in a hospital in Peru. Rev. Peruana Med. Exp. Salud Pública **39**(4), 415–24 (2022). https://doi.org/10.17843/rpmesp.2022.394.11287
14. Vargas González, V., Hernández, C.: Sistemas de información de costos para la gestión hospitalaria. Rev. Cienc. Sociales **15**(4) (2009). https://doi.org/10.31876/rcs.v15i4.25477
15. Valeri Ramírez, L.C.: Los sistemas de información para la gerencia en salud pública. Vis. Gerencial **2**, 435–460 (2016). http://erevistas.saber.ula.ve/index.php/visiongerencial/article/view/7749
16. Riveros, J., Berné, C., Múgica, J.: Gestión y satisfacción en servicios de salud de chile: contraste entre las percepciones de los funcionarios y los usuarios. Rev. Med. Chil. **138**(5), 630–638 (2010). https://doi.org/10.4067/S0034-98872010000500015
17. Rodríguez, D.C., et al.: Módulo consulta externa del sistema de información hospitalaria xavia his. Rev. Cubana Inf. Méd. **14**(2), e540 (2022). https://revinformatica.sld.cu/index.php/rcim/article/view/540
18. Bocquet, F., Raimbourg, J., Bigot, F., Simmet, V., Campone, M., Frenel, J.S.: Opportunities and obstacles to the development of health data warehouses in hospitals in France: the recent experience of comprehensive cancer centers. Int. J. Environ. Res. Pub. Health **20**(2) (2023). https://doi.org/10.3390/ijerph20021645
19. Granda, W.X.B., Molina-Granja, F., Altamirano, J.D., Lopez, M.P., Sureshkumar, S., Swaminathan, J.N.: Data analytics for healthcare institutions: a data warehouse model proposal. In: Inventive Communication and Computational Technologies, pp. 155–163. Springer, Singapore (2023). https://doi.org/10.1007/978-981-19-4960-9_13
20. Inmon, W.H.: Building the Data Warehouse, 4th edn. Wiley (2005)
21. Linstedt, D., Olschimke, M.: Building a Scalable Data Warehouse with Data Vault 2.0. Morgan Kaufmann (2015)
22. Duran, A.: Un entorno metodológico de ingeniería de requisitos para sistemas de información, Ph.D. thesis, Universidad de Sevilla (2000). http://hdl.handle.net/11441/15365

School Dropout and Regional Competitiveness in Peru

Sebastián Vilca-Porras(✉), Alex Manrique-Peralta, Joaquín Alonso Ortega-DelAguila, Alvaro Salazar Gonzales, and Victor Andrés Ayma Quirita

Universidad del Pacífico, Lima, Peru
{se.vilcap,aj.manriquepe,ja.ortegad,a.salazargo}@alum.up.edu.pe, va.aymaq@up.edu.pe

Abstract. School dropout in Peru is a highly significant phenomenon that directly affects the productivity and competitiveness of society. This research explores the relationship between school dropout and regional development in Peru in 2023, using data from the National Household Survey (ENAHO) and the Regional Competitiveness Index (INCORE). Analytical methods, including multiple linear regression, the regression equation-specific error test, and principal component analysis (PCA), were employed to examine how school dropout correlates with regional competitiveness. The results reveal a negative association between regional development and the school dropout rate. Regions with lower competitiveness, particularly in the jungle and high Andean areas, exhibit the highest dropout rates, while regions with high competitiveness, such as Moquegua and Arequipa, show significantly lower rates. These findings highlight the need for targeted public policies to improve educational quality and reduce regional inequalities, decrease school dropout, and promote balanced social and economic development throughout the country.

Keywords: school dropout · regional competitiveness · education · multiple linear regression · principal component analysis · regression equation specification error test

1 Introduction

The phenomenon of school dropout is generally studied internationally due to its relevance in the context of economic development for future generations around the world. In Peru, studies such as those of Alarcón [1], Fernandez and Rosales [6], and Lavado and Gallegos [10] have already worked on the multivariate analysis of school dropout in Peru, taking school dropout as a dependent variable and relating it to other variables such as mother tongue, local area where they live, parents' education, and region where they live. However, these studies might not account for recent changes in these variables.

F. Suni-Lopez et al. (Eds.): CIIS 2024, ISEM 48, pp. 91–105, 2025.
https://doi.org/10.1007/978-3-031-91481-2_6

Authors such as Lavado and Gallegos [10] define dropouts as those people who, having attended school the previous year, do not do so in the current year, excluding those who have stopped attending because they have completed secondary school, are studying at a pre-university academy, were ill, had an accident, or are doing their military service. In this way, the development of the research on school dropout in Peru can be initially proposed, considering that the information necessary to carry out the study is abundant and granular for each case. However, despite the lack of data to calculate the dropout index per region, we propose the use of some of the variables related to school dropout for the study.

The dynamics of school dropout are generally discussed due to the role that this phenomenon plays in Peru's economic development. Muñoz de Bustios et al. [11] mention that early school dropout is related to the possibility of changing school for work. In this way, it is mentioned that the group of young people who leave secondary education prematurely is at greater risk of unemployment and inactivity, while they may occupy unskilled and low-wage jobs. As already explained by Choudhary et al. [4], dropout cases are influenced by various factors around the world, which makes it important to analyze these variables specifically for Peru.

The main motivation of this research is to explore the relationship between the dropout rate in schools and the regional development (interpreted by the regional competitiveness index). This aims to understand and empirically substantiate the direct consequences of school dropout on the economic and social development of a society. We aim to explore the relationship between school dropout and regional competitiveness in Peru in the year 2023. In addition, we seek to contribute to the debate on the need for public policies consistent with the reality of each region that reduce disparities and promote social welfare.

Currently, the literature presents various studies on school dropout in Peru, the associated factors, and its possible effects. A first analysis proposed by Lavado and Gallegos [10] establishes that economic restrictions related to family poverty are a significant determinant of school dropout. Then Alcazar [2] adds that the relationship between poverty and the decision not to continue attending school is more complex, stating associated factors such as the evaluation of education and particular learning or family problems.

In addition, Alcazar [2] highlights that the region of residence is related to the decision to study by paying the opportunity costs of this activity. Consistent with this, Gutierrez [8] explains that infrastructure, such as school connectivity, is crucial for continuing education, especially after COVID-19. However, Reátegui [13] explains that school closure during the state of emergency could have generated a 21.3% decrease in years of schooling, impacting student quality of life and long-term financial prospects.

Although the previous literature includes an analysis of associated factors and possible effects of school dropout, from a national perspective, the present study aims to provide a regional approach to school dropout and its relationship with regional competitiveness and economic development.

2 Related Works

According to Castro [3], dropping out of school leads to a reduction in the job opportunities of an individual because they lack the knowledge and skills necessary to meet the demands of work. As a result, students who drop out of school are candidates to become vulnerable populations, which generates high private and social costs. Thus, in the long term, higher social costs are generated in many regions of Peru. The consequences of these events are reflected in the Peruvian Regional Competitiveness Index (INCORE), which will be discussed later [12].

Furthermore, Cueto's [5] analysis, explains that the earlier the dropout occurs, the greater the effect observed on skills later in the student's life. Mainly, the cause is the need to work to earn money, due to the family's monetary conditions. Another factor in school dropout is performance during primary school. Poor academic achievement in that period might increase the dropout risk; this also covers grade repetition.

Regarding the reference framework, this study employs a simple linear regression analysis that considers the INCORE variable as the independent variable and the school dropout as the dependent variable to establish relevant conclusions and generate graphic representations to support the statements. Likewise, a simple analysis of the percentage of students who dropped out of school and INCORE is used to obtain graphical representations, such as the heat map.

3 Theoretical Framework

3.1 Regional Competitiveness

Regional development is defined and quantified through INCORE, an indicator that evaluates the performance of metropolitan Lima, Callao, and the 24 regions of the country through six pillars: economic environment, labor, health, education, institutions, and infrastructure [9]. Thus, the information is used to determine the level of competitiveness of Peru's regions.

3.2 School Dropout

In first place, school dropout is defined as the "conditions under which failed integration occurs in the academic and social environment of institutions" [16] and can be quantified by counting the "students who were enrolled in year t and who do not appear to be enrolled in year t+1, excluding those who died in year t or completed secondary education" [7]. In this case, school dropout will be our variable X, looking for a possible relationship with regional competitiveness.

3.3 Multiple Linear Regression

Wooldridge [17] explains that multiple regression analysis is more suitable for a ceteris paribus analysis because it allows for the explicit control of many other factors that simultaneously affect the dependent variable, as shown in the model of Eq. 1.

$$y = \beta_0 + \sum_{i=1}^{n} \beta_i x_i + \epsilon \tag{1}$$

where:

- y: Dependent variable.
- β_0: Intercept term, representing the expected value of y when all x_i are zero.
- β_i: Coefficient for x_i, for $i = 1, 2, \ldots, n$.
- x_i: Independent variables
- ϵ: Error term, which captures the unexplained variability in y

Furthermore, an ordinary least squares (OLS) estimate is used for this analysis to minimize the sum of squared residuals. That is, given n observations on y, x_1, and x_2, $\{(x_{i1}, x_{i2}, y_i) : i = 1, 2, \ldots, n\}$, the estimates $\hat{\beta}_0$, $\hat{\beta}_j$ are chosen simultaneously to minimize the objective function in Eq. 2.

$$\text{Minimize} \quad \sum_{i=1}^{n} \left(y_i - \hat{\beta}_0 - \sum_{j=1}^{p} \hat{\beta}_j x_{ij} \right)^2 \tag{2}$$

3.4 Regression Equation Specification Error (RESET) Test

Wooldridge [17] proposes the regression equation specification error (RESET) test as a method to detect incorrect specifications of functional forms. For this test, squares of significant explanatory variables were added to test the linearity of the functional form of the regression. The null hypothesis is that the functional form is correctly specified (linear). Thus, the F statistic is used to test $H_0 : \gamma_1 = 0, \gamma_2 = 0$ in the expanded model as shown in Eq. 3. A significant F statistic suggests some form of functional form issue.

$$y = \beta_0 + \sum_{j=1}^{k} \beta_j x_j + \gamma_1 \hat{y}^2 + \gamma_2 \hat{y}^3 + \epsilon \tag{3}$$

where:

- y: Dependent variable.
- β_0: Intercept term.
- $\sum_{j=1}^{k} \beta_j x_j$: Sum of the original independent variables x_j multiplied by their coefficients β_j, where $j = 1, 2, \ldots, k$.
- $\gamma_1 \hat{y}^2$ and $\gamma_2 \hat{y}^3$: Terms for the squared and cubed fitted values, testing for potential nonlinearities in the model.
- ϵ: Error term, which captures the unexplained variability in y.

3.5 Principal Component Analysis (PCA)

Principal component analysis (PCA) is a mathematical technique that transforms a set of possibly correlated variables into a smaller set of variables known as principal components. The technique originated in multivariate data analysis and is widely used in applications such as dimensionality reduction of large datasets, noise elimination in signals, source separation, and data compression. It reduces data dimensionality by transforming the vector space, allowing the original data to be represented with fewer variables, thus facilitating the detection of trends, patterns, and outliers [14].

4 Materials and Methods

In the present work, the use of sources related to the educational topic was determined to fulfill the objective of this investigation. In this section, we detail exactly what these materials were, followed by the programming tools, theory, and reports that contributed greatly to this study on school dropout.

The flow chart depicted in Fig. 1 explains the procedure of our approach in a structured way, which concisely shows the collection of bases for the analysis of the variables. The first step involves registering the INCORE and ENAHO data frames, followed by filtering these data frames with the variables of interest. Subsequently, a new binary variable of school dropout is created. The next step is to merge the INCORE and National Household Survey (ENAHO) databases into one data frame. After merging, the proportion, total, and number of dropouts for each region are calculated. A merge of ENAHO modules 3 and 4 is also performed to create a final data frame. Using this final data frame, heat maps are created to show the percentage of dropouts in each region. Similarly, heat maps are created with INCORE results per region. Following this, a regression with the corresponding variables is executed, and the results of these regressions are collected and interpreted. Furthermore, the analysis is conducted through PCA. Finally, a RESET test is conducted to analyze the linearity function.

4.1 Data Collection

The ENAHO is the main source of this work, as, based on these surveys, it is possible to collect information on the educational environment of families in a provincial sector. That said, the base collected is from modules 3 and 4 of the ENAHO, in the year 2023.

Similarly, along with the ENAHO, another major source was the INCORE 2023 [9], which provides a result that summarizes the six most important competitiveness pillars: economic environment (P1), labor (P2), health (P3), education (P4), institutions (P5), and infrastructure (P6). Thus, by placing the results for the year 2023 per region and pillar, a data frame of INCORE is obtained.

To create the databases for this work, first we needed to import the previously mentioned databases and convert them into Data Frames. In the case of ENAHO,

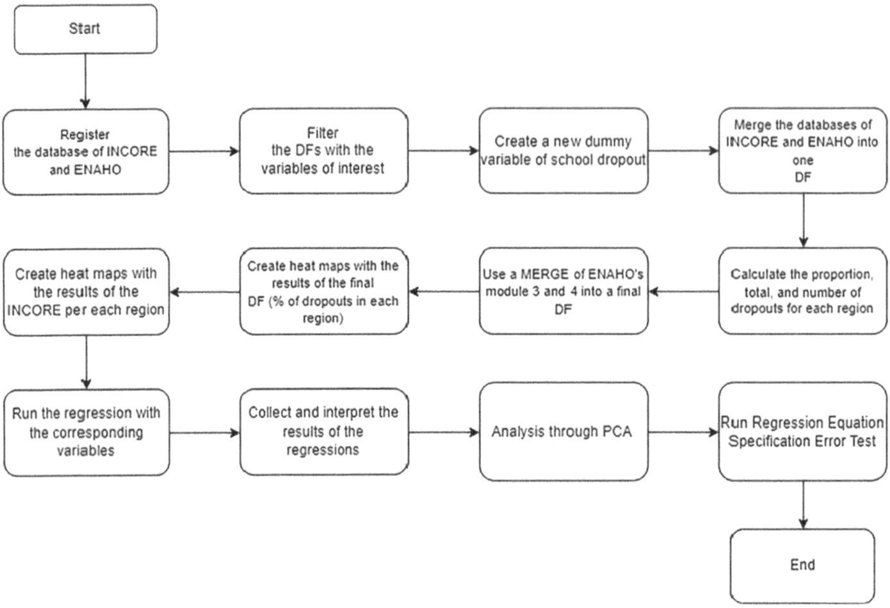

Fig. 1. Methodology

to start with the modifications of the variables, as well as the use of filters, it was necessary to have a dictionary provided by ENAHO as an aid, which kept a record of the meaning of all the variables and their answers in numerical format. This made it possible to filter the database, so we finally had variables of interest based on the provinces of Peru.

For this case, modules 3 and 4 of the ENAHO were used to obtain the variables necessary to create the binary variable (dummy) of the dropout from the school, as shown in Table 1 and Table 2, where the variables selected and conditional for the creation of the dropout variable from the school are explained. For this case, the variables P301A, month, and age were fundamental in setting a strict parameter for the dummy variable of school dropout. For example, as shown in Table 2, the variable P301A was delimited by cases where it is 3 and 5 as it represents incomplete primary and secondary school studies.

4.2 Data Processing

The INCORE case presented significant differences, primarily involving the modification of column and row names to enhance understanding. Furthermore, the school dropout rate was derived from data collected previously, using the criteria described in Table 2.

Finally, after data processing, the product includes a final data frame in which the school dropout ratio and the INCORE pillars are matched per province[1].

[1] Repository in github.

Table 1. Classification of selected variables from ENAHO

Variable	Description	Type of variable
P301A	What is the highest grade or level of education you have completed?	Categorical
Month	The month when the survey has been done	Quantitative
P400A2	What day, month, and year were you born? - Month	Quantitative
P400A3	What day, month, and year were you born? - Year	Quantitative

Table 2. Conditionals for binary (dummy) variable of school dropout

Variable	Conditional
P301A	`==3 (for primary school)` `==5 (for secondary school)`
Month	`>=4`
Age by June (Created using P400A2 and P4003)	`>=7 and <=14 (for primary school)` `>=13 and <=19 (for secondary school)`

4.3 Multiple Linear Regression Analysis

With the previous data frame implemented, a multiple linear regression was performed, considering the competitiveness index as the independent variable and the regional school dropout rate as the dependent variable. This analysis aims to explore the causal relationship between school dropout rates and productivity levels in the regions of Peru. For this multiple regression, we used the different pillars of the INCORE and the general INCORE index for each region. Each coefficient β_i quantifies the approximate change in the regional dropout rate of school (y) associated with a change of one unit in the corresponding variable x_i, assuming that all other variables remain constant. Furthermore, to test for the function's linearity, we use a RESET test as explained before. This approach ensures that the viability of the OLS method is optimal for our analysis.

4.4 Principal Component Analysis (PCA) Approach

A multicollinearity analysis was performed to adjust the model by examining the regression correlation matrix of the independent variables, considering that high correlations between independent variables can lead to multicollinearity, which inflates the variance of the coefficient estimates and makes the model unstable. To mitigate this issue, PCA was applied to the seven independent variables, and it was determined, after evaluating the criterion of the eigenvalue greater than 1, that only one principal component should be maintained, which captures the greatest amount of total variability of the data. Finally, with this principal component, which represents the INCORE, since it is constructed with the six pillars evaluated by the index, a simple linear regression was performed where the dependent variable is the percentage of population attrition and the independent variable is the principal component generated.

4.5 Limitations of the Study

Finally, to further expand the analysis, we developed a series of heat maps reflecting school dropout at the provincial level, so that it would be visible in which provinces it is more recurrent, and a heat map regarding the INCORE. In this way, the most essential results of this research were obtained. However, it is worth mentioning the following points:

- The ENAHO database does not permit us to know exactly the cases of school dropout, thus an approximation is made through the creation of the new dummy variable of school dropout.
- Being a database that bases its information on surveys, the possibility of deception and false information is open, which represents a lack of reliability of the ENAHO database, but not of the INCORE.
- Finally, changes in regional competitiveness and school dropout rates may exhibit spatial and temporal dynamics that the ENAHO cannot fully capture, potentially limiting the accuracy of long-term trend analyses, which could have an intertemporal effect.

5 Results

Through the application of the previously mentioned tools and methods, the results derived from the elaboration process of this study are presented through a series of graphs and tables. The resources obtained are vital to draw pertinent conclusions about the objectives outlined in this investigation.

The methods used in the study produced various graphs and tables crucial for drawing subsequent conclusions. First, Fig. 3 (right) shows the heat map of the INCORE for the year 2023. Second, Fig. 2 (left) presents the heat map showing the percentage of school dropouts by region in Peru during the same period. Following these, Figs. 4 and 7 depict the regression reports generated by Python for dropout 2023 and INCORE 2023, as well as the PCA. Figure 5 illustrates the results of the regression analysis of school dropout with the education pillar, which was found to be the only significant pillar. Finally, Fig. 6 shows the correlation matrix of the independent variables used in the regression model to analyze multicollinearity.

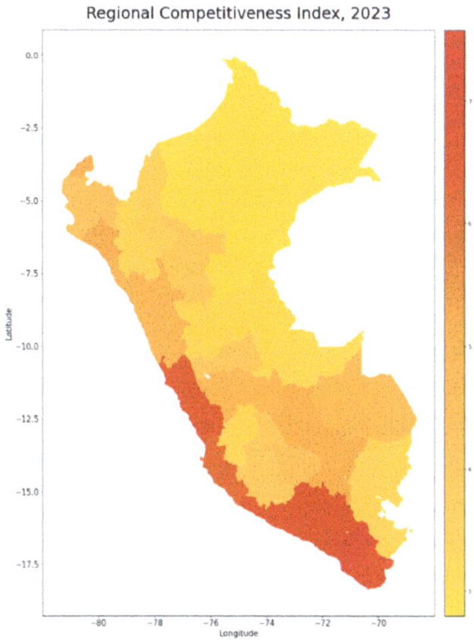

Fig. 2. Heat map of Regional Competitiveness Index of Peru, 2023

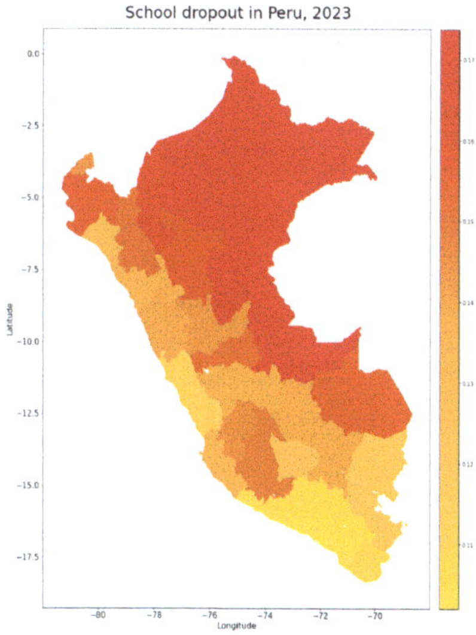

Fig. 3. Heat map of regional school dropout in Peru, 2023

Figure 2 reveals that Moquegua has the highest INCORE score. In addition, coastal departments, especially those in the south, are observed to exhibit the highest INCORE score, in contrast to highland and jungle departments, which show considerably lower rates. Similarly, Fig. 3 shows that Loreto has the highest dropout rate compared to other departments. In contrast, both Moquegua and Arequipa present the lowest school dropout rates.

Concerning the regression report presented in Fig. 4, results such as the determination coefficient (R2) stand out, which exhibits a regular explanation (70.7%) of school dropout by the independent variable. Likewise, this case can also be explained through the P-value in both cases where it is less than 0.05, which can be translated into a high significance of the same within the regression for the dropout dependent variable for the year that has been studied. In this case, the independent variable of education reflects significance for the regression. This means that, the pillar of education in INCORE has a negative relationship with school dropout as shown in Fig. 5. Therefore, as the education pillar in the INCORE decreases, school dropout rates tend to increase.

Moreover, to analyze cases of multicollinearity, Fig. 6 displays the correlation of the regression's independent variables. The results show that there is a moderate and strong correlation between most of the independent variables. Therefore, the PCA methodology is proposed to mitigate this problem. After its application, a new principal component was obtained that explains 78.24% of the variance of the original independent variables. Subsequently, an OLS regression was performed to evaluate the relationship between regional school dropout and

```
                            OLS Regression Results
==============================================================================
Dep. Variable:        Proporci�n_2023   R-squared:                       0.783
Model:                            OLS   Adj. R-squared:                  0.707
Method:                 Least Squares   F-statistic:                     10.23
Date:                Wed, 26 Jun 2024   Prob (F-statistic):           7.39e-05
Time:                        23:01:06   Log-Likelihood:                 78.135
No. Observations:                  24   AIC:                            -142.3
Df Residuals:                      17   BIC:                            -134.0
Df Model:                           6
Covariance Type:            nonrobust
================================================================================
========
                           coef    std err          t      P>|t|      [0.025
0.975]
--------------------------------------------------------------------------------
--------
const                    0.1633      0.017      9.361      0.000       0.127
0.200
Economic_environment    -0.0006      0.004     -0.156      0.878      -0.008
0.007
Infrastructure          -0.0003      0.004     -0.077      0.940      -0.009
0.008
Health                  -0.0012      0.003     -0.455      0.655      -0.007
0.004
Education               -0.0096      0.003     -3.473      0.003      -0.015
-0.004
Labor                    0.0037      0.003      1.049      0.309      -0.004
0.011
Institutions             0.0039      0.004      1.091      0.291      -0.004
0.012
General                 -0.0007      0.000     -1.381      0.185      -0.002
0.000
==============================================================================
Omnibus:                        0.972   Durbin-Watson:                   2.080
Prob(Omnibus):                  0.615   Jarque-Bera (JB):                0.883
Skew:                          -0.417   Prob(JB):                        0.643
Kurtosis:                       2.567   Cond. No.                     2.65e+16
==============================================================================
```

Fig. 4. OLS regression results

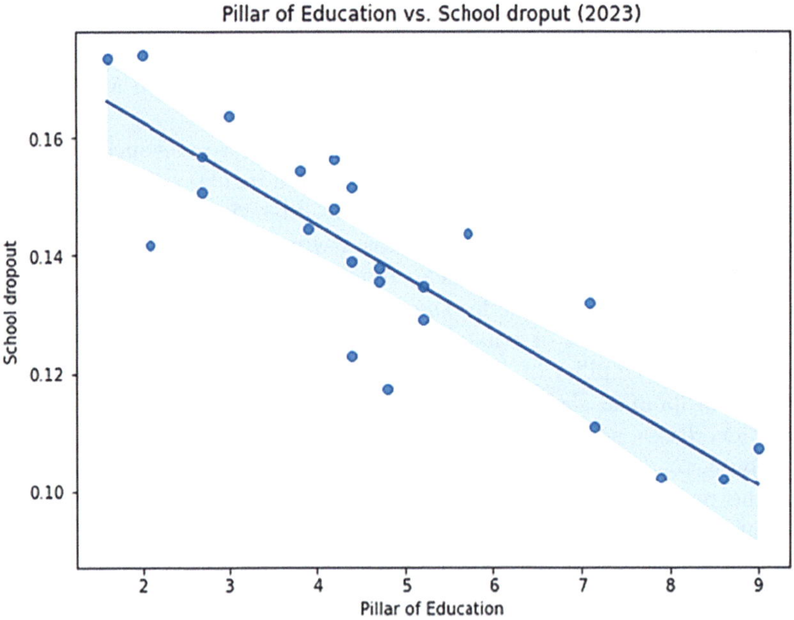

Fig. 5. School dropout vs. education regression chart

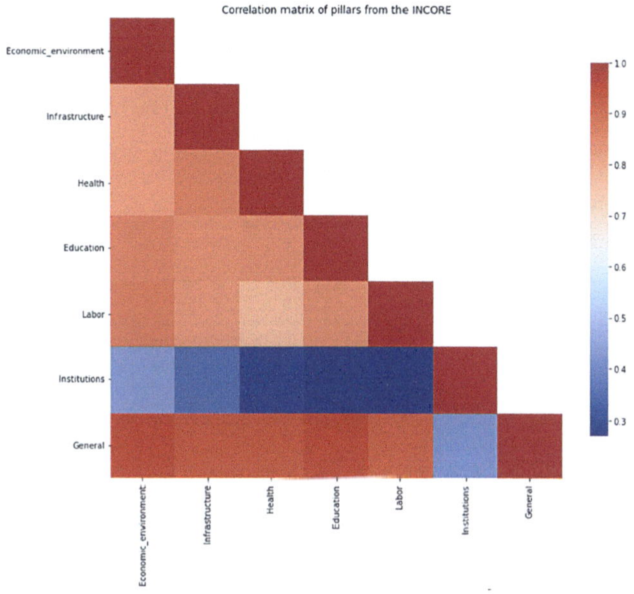

Fig. 6. Correlation Matrix of the pillars from the INCORE

the identified principal component. The results indicate a negative relationship between the dependent and independent variables; that is, the lower the regional development, the higher the dropout rate. Regarding the adequacy of the model, there is evidence of significance both individually, concerning the independent variable, and globally, concerning the implemented model. In addition, the R^2 is 60.6%, as shown in the Fig. 7. Moreover, to ensure the appropriateness of the applied method, we conducted a RESET test, which resulted in a P-value of 0.89. The model tends to exhibit linearity since this value is higher than 0.05.

5.1 Discussion

One of the most notable aspects revealed by this study lies in the observation in Fig. 2, where the contrast in the levels of development between the different departments is highlighted. The main example of this disparity is the capital city, Lima, which is significantly different from other departments as a whole. However, when comparing these data with Fig. 3, it is evident that the dropout rate in Lima is not necessarily the lowest, which contradicts a possible expectation based on the level of development.

This phenomenon is perfectly explained by reviewing the results of the regression report presented in Fig. 4, which shows the low relationship between the variables. The opposite case can be observed in the jungle and highlands areas in the south of the country where low INCORE levels can be seen and the observation of school dropout is higher, which shows a negative relationship between these cases.

OLS Regression Results

Dep. Variable:	Proporción_2023	R-squared:	0.606
Model:	OLS	Adj. R-squared:	0.588
Method:	Least Squares	F-statistic:	33.87
Date:	Fri, 28 Jun 2024	Prob (F-statistic):	7.43e-06
Time:	12:44:29	Log-Likelihood:	70.978
No. Observations:	24	AIC:	-138.0
Df Residuals:	22	BIC:	-135.6
Df Model:	1		
Covariance Type:	nonrobust		

	coef	std err	t	P>|t|	[0.025	0.975]
const	0.1387	0.003	51.745	0.000	0.133	0.144
PC1	-0.0067	0.001	-5.820	0.000	-0.009	-0.004

Omnibus:	6.931	Durbin-Watson:	2.081
Prob(Omnibus):	0.031	Jarque-Bera (JB):	4.754
Skew:	-0.960	Prob(JB):	0.0928
Kurtosis:	4.034	Cond. No.	2.34

Fig. 7. PCA Analysis Results

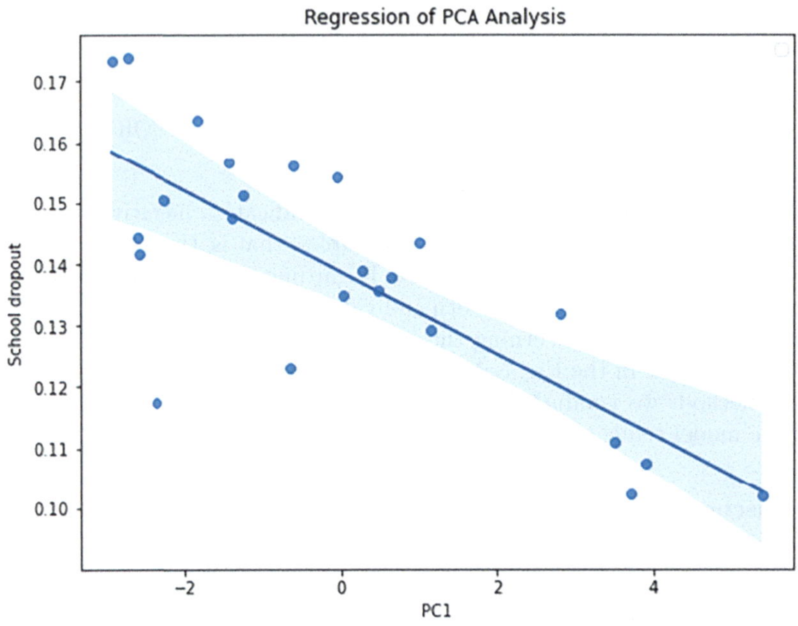

Fig. 8. School dropout vs. PCA regression chart

Based on the findings, we have been able to satisfactorily identify the departments with the highest levels of school dropout at the national level. In addition, we were able to make a comparison between these departments using a heat map. However, the data collected from Fig. 4 indicate an almost null explanation between the variables studied: regional development according to the INCORE and the rate of school dropout.

As previous studies suggest, there may be a relationship between school dropout due to the substitution of studies for jobs, given the quality of life in their region. It is true that, through the regressions proposed, the relationship of the INCORE with dropout can be dismissed; however, through the graphs, it can be observed that in certain regions this hypothesis may be close to reality.

6 Conclusions

The present research explored the relationship between school dropout and regional competitiveness in Peru, evaluating private and social costs, and the impact on the productivity and competitiveness of the regions.

The results of the regression analysis established that there is no relevant relationship between all the variables studied, possibly due to the high variability of the student dropout sample. However, by disaggregating the data through the graphical representation of the heat map, trends and patterns were identified that allow a better understanding of the complexity of the phenomenon.

For example, the case of many regions in the jungle (Loreto, Ucayali, Amazonas, San Martín) presents a negative relationship between INCORE and the rate of school dropout. As seen, the general competitiveness index could lead to a higher school dropout rate due to the quality of the education service and the factors related to this pillar, such as access to the internet, basic services, and student performance in the region.

This implies first the need to rethink the collection of statistical information related to school dropouts in national surveys, so that it provides more relevant information. Secondly, following the analysis and interpretation of the graphical representations of the heat map, it implies the need for targeted public policies such as dropout prevention programs, educational reinsertion programs, and financial and resource support to address highly vulnerable populations.

Based on the results, there is the possibility of extending the analysis carried out through an intertemporal regression that evaluates the medium- or long-term impact of the dropout decision, to make the proposal more consistent. Likewise, there is the possibility of modifying the focus from the consequences of dropping out to the causes of dropping out, a study that would make it possible to define the best strategies to follow.

6.1 Recommendations

As stated above, our results provide interesting insight and recommendations. First, competent authorities and non-governmental organizations should actively intervene to raise awareness of the long-term impacts of school dropout. This intervention is crucial both to improve individual career prospects and to increase social productivity in general. According to the previous assessment conducted, there is a clear correlation: the lower the development and competitiveness, the higher the dropout rate. This results in a less skilled and less productive population, which in turn perpetuates a vicious cycle of lower development and competitiveness.

Secondly, regional competitiveness conditions must be improved to have a positive impact on the reduction of the school dropout rate; that is, public policies must be aimed at guaranteeing equitable access to quality education for all sectors of the population, with special attention to vulnerable groups and marginalized communities, through a better distribution of public resources.

Finally, for future related research, we suggest studying the causes of school dropout at the family or household level so that the main difficulties faced by students in their immediate environment can be identified with greater precision. This could include the analysis of factors such as the economic situation, the educational level of parents, family dynamics, and the availability of resources and support at home. Furthermore, we encourage future research using a more granular competitiveness index that includes both towns and districts, allowing for more representative studies. By delving deeper into these aspects, it is hoped to develop more effective and targeted interventions to reduce dropout, improving student retention and academic success.

References

1. Alarcon, W.: Atraso y deserción escolar en niños y adolescentes. Dirección Técnica de Demografía y Estudios Sociales del INEI, Lima (1995). https://proyectos.inei.gob.pe/web/biblioineipub/bancopub/est/lib0023/N00.htm
2. Alcázar, L.: Asistencia y deserción en escuelas secundarias rurales del Perú. In: Benavides, M. (ed.) Análisis de programas, procesos y resultados educativos en el Perú: Contribuciones empíricas para el debate. GRADE (2008). https://www.grade.org.pe/wp-content/uploads/analisis-1.pdf
3. Castro, M.: Estrategia de sensibilización para evitar la deserción escolar de niñas y niños de nivel primario a través del proyecto Pienso, opino y participo de Global Humanitaria Perú, Master's thesis, Universidad de Lima. Institutional Repository of the Universidad de Lima (2021). https://repositorio.ulima.edu.pe/handle/20.500.12724/13741
4. Choudhary, L.A., Latif, A., Hammayun, A.A.: Economic effects of student dropouts: a comparative study. J. Glob. Econ. **3**(2), 1–6 (2015). https://doi.org/10.4172/2375-4389.1000137
5. Cueto, S., Felipe, C., León, J.: Predictores de la deserción escolar en el Perú. Análisis y Propuestas, No. 52. GRADE, Lima (2020). https://www.grade.org.pe/publicaciones/predictores-de-la-desercion-escolar-en-el-peru/
6. Fernandez, H., Rosales, J.: Educación, Una Mirada Hacia Adentro: Analfabetismo, Repitencia y Deserción. Instituto de Pedagogía Popular, Lima (1990)
7. García, D., et al.: Contribución del proyecto aula hospitalaria a la permanencia en el sistema educativo de estudiantes hospitalizados. Tesis de maestría, Universidad del Pacífico. Repositorio Institucional de la Universidad del Pacífico (2021). https://repositorio.up.edu.pe/handle/11354/2918

8. Gutiérrez, J.: Mayor conectividad entre escuelas reduce la deserción escolar? Un análisis geoespacial en la transición educativa primaria-secundaria en el Perú, Tesis de maestría, Universidad del Pacífico. Repositorio Institucional de la Universidad del Pacífico (2022). https://repositorio.up.edu.pe/handle/11354/3393
9. Instituto Peruano de Economía: Índice de Competitividad Regional - INCORE 2023. Lima, Perú (2023). https://incoreperu.pe/portal/index.php/ediciones-anteriores/item/17-incore-edicion-2023
10. Lavado, P., Gallegos, J.: La dinámica de la deserción escolar en el Perú: un enfoque usando modelos de duración. Ministerio de Educación del Perú (2005). https://repositorio.minedu.gob.pe/handle/20.500.12799/253
11. Muñoz de Bustillo, R., et al.: Abandono escolar y mercado de trabajo en España. (2009). https://www.researchgate.net/publication/304715906_Abandono_escolar_y_mercado_de_trabajo
12. Ramsdal, G.H., Wynn, R.: Attachment and school completion: understanding young people who have dropped out of high school and important factors in their re-enrollment. Int. J. Environ. Res. Public Health **19**(7), 3938 (2022). https://doi.org/10.3390/ijerph19073938
13. Reátegui, L.: Pandemia y deserción escolar en la educación básica regular: factores asociados y posibles efectos, 2017–2021 (2022). https://www.inei.gob.pe/media/MenuRecursivo/investigaciones/desercion-escolar.pdf
14. Richardson, M.: Principal component analysis (2006). http://www.sdss.jhu.edu/~szalay/class/2024/etc/SignalProcPCA.pdf
15. Román, M. C.: Factores asociados al abandono y la deserción escolar en América Latina: una mirada en conjunto. REICE Rev. Iberoam. sobre Calid. Efic. y Cambio en Educ. **11**(2), 33–59 (2013). https://www.redalyc.org/pdf/551/55127024002.pdf
16. Tinto, V.: Una consideración de las teorías de la deserción estudiantil en la trayectoria escolar en la educación superior. ANUIES, México (1987)
17. Wooldridge, J.M.: Introductory Econometrics: A Modern Approach, 7th edn. Cengage Learning, USA (2020)

Tagging with *pysentimiento* and its Evaluation in Traditional Supervised Learning Models for Sentiment Analysis

Claudia Francesca Suarez Mariscal, Yeshica Isela Ormeño Ayala(✉), Jose Mauro Pillco Quispe, and Lino Prisciliano Flores Pacheco

Departamento Académico de Ingeniería Informática y de Sistemas, Universidad Nacional de San Antonio Abad del Cusco, Cusco, Perú
{claudia.suarez,yeshica.ormeno,jose.pillco,lino.flores}@unsaac.edu.pe

Abstract. The excessive growth of opinions or emotions in different social networks has enabled the optimization of sentiment analysis (SA). However, the data labeling phase usually takes a lot of time and resources if manual labeling is performed, due to the amount of data. The present study seeks to use bidirectional transformer models (*pysentimiento*) to label tweets in Spanish, considering political and health events in Peru. For this purpose, a methodology for the construction of a dataset is proposed, which includes the extraction of tweets by Twint, the filtering and cleaning of data, the labeling of tweets by *pysentimiento*, the balancing of classes, and the manual verification of tweets. Through the experimental phase, where traditional supervised models-naive Bayes (NB), super vector machine (SVM), random forest (RF), and multilayer perceptron (MP)-are trained with the proposed datasets, the improvement in the performance of each model is shown by cross validation and the stratification of classes. The results show that manual verification contributes to an optimal classification of tweets in the positive and negative categories, but not in the neutral category.

Keywords: Analysis of feeling · *twint* · *Pysentimiento* · Machine Learning

1 Introduction

In social networks, the opinion of each user is one of the central aspects with which different behaviors could be identified. Interaction between users and social networks is increasing. *Twitter*, for example, hosts the views of millions of users on real-time events [9]. Sentiment analysis of tweets (SAT) is positioned as a research approach in social networks oriented to the study of opinions, attitudes, values, or emotions towards different products, events, organizations, themes, or characters, etc. [12]. However, SAT is not a trivial task as it deals with different types of information contained in a tweet (hyperlinks or emoticons), as well as the volatility of language or paraphrasing [7].

F. Suni-Lopez et al. (Eds.): CIIS 2024, ISEM 48, pp. 106–120, 2025.
https://doi.org/10.1007/978-3-031-91481-2_7

In data science, it is important to count or select an appropriate dataset for the learning process [15]. When collecting tweets is required, Twitter provides its data (tweets) through APIs, but only a limited amount of data can be accessed. A raw data labeling step requires more time and resources, since it is performed manually [2], however, automatic labeling can be used through specialized models for text classification such as the bidirectional encoder representations from transformers (BERT) model [6].

There are several studies that show the effectiveness of BERT models, but most of the analyses are carried out for the English language [5], as to use it in another language, it is necessary to determine which versions of BERT could work better. Recent studies [22,25] show that pysentimiento, a BERT variation for the Spanish language, is a versatile tool for classifying tweets in positive, negative, and neutral according to the polarity score. However, most studies only compare the performance of the model with other classifiers [5,12,17,24] and do not consider the efficiency in other purposes, such as during the labeling phase of tweets.

The motivation for the present study is to apply pysentimiento in the automatic labeling phase through traditional supervised models. Based on this premise, we seek to i) extract tweets without limitations using Twitter web-scraping techniques; ii) reduce labeling time using pysentimiento; iii) evaluate the efficiency of pysentimiento using four traditional classification algorithms; and iv) optimize the performance of each model in the training and testing phase.

The present study is structured as follows: Sect. 2 describes the studies related to the present study proposal; Sect. 3 presents the methodology used to construct the dataset; and Sect. 4 presents the four experiments to improve the performance of the supervised learning models. Finally, Sect. 5 describes the conclusions.

2 Related Studies

This section includes studies focused on the evaluation of bidirectional transformer-BERT-models, giving greater emphasis to pysentimiento or other tools for the Spanish language.

Miranda et al. [14] proposed evaluating the feelings in tweets in Spanish about the pandemic. Web scraping techniques were used in data extraction, and preprocessing was applied to filter and clean the data; then BERT-base was used to perform sentiment classification. Arias et al. [1] carried out a sentiment analysis of tweets in Panama during the years 2019 and 2020, and highlighted the use of natural language processing (NLP) to consider linguistic regionalisms through BERT. In this way, BERT can be considered as a configurable tool in different languages that deals with language-specific characteristics such as slang or regionalisms; however it still has imperfections since it is based on the English language.

For BERT models specialized in Spanish, there is pysentimiento [19]. Pérez et al. [20] analyzed news headlines related to the 2019 Argentine presidential elections. Classification algorithms based on transformer models such as BETO [3], RoBERTaes [21], BERTin [23] and RoBERTuito [18] were used. For this, pysentimiento provided the dataset TASS 2020 for the training phase. Prathap et al. [17] built a piece of software to identify the anti-American sentiment in fake news using pysentimiento and Text2Emotion. In another context, Rateb et al. [22] used convolutional neural network-long short-term memory (CNN-LSTM), super vector machine (SVM), and pysentimiento for the analysis of feelings in tweets about cryptocurrencies (translated into Spanish) during the Russian-Ukrainian war. Gomez-Adorno et al. [8] evaluated machine learning and deep learning models (together) for the analysis of feelings in Twitter SENT-COVID using TextBlob, VADER, and pysentimiento. Among these studies, pysentimiento showed superiority in classifying Spanish text, as demonstrated by Sahoo et al. [24] by means of a comparative analysis between BERT uncased and pysentimiento, with the latter performing better.

Based on the analysis of the studies carried out, it is observed that pysentimiento is an appropriate tool to identify feelings in tweets in Spanish, through its three analyzers: feeling, emotion, and hate speech [11]. The present study seeks to evaluate such efficiency in another way, specifically in using pysentimiento for automatic labeling of raw data.

3 Methodology

This section presents the description of the stages of the proposed methodology (Fig. 1) for the construction of the dataset, which includes data acquisition by Twint, data cleaning, breakdown of tweets, tweet labeling using pysentimiento, data balancing by category, and finally, a manual verification. As a result, two datasets were obtained: one in the political context, and the other in the pandemic context. Additionally, data science processes were followed to use these datasets in the training of traditional supervised learning models, which were naive Bayes (NB), super vector machine (SVM), random forest (RF), and multilayer perceptron (MP).

3.1 Construction of the Dataset

1. **Extraction of tweets with Twint:** In this phase, tweets were extracted using the web scraping tool Twint, using hashtags and usernames. For extraction with hashtags, 89 hashtags were used in the political context, 115 in the pandemic context, and 14 usernames related to each context. The list of hashtags and usernames was obtained by searching for trending topics on Twitter in Peru on getdaytrends[1]. Some of the selected hashtags were #KeikoFujimori, #PedroCastillo, #pandemia, #vacunacion, and #Covid19. The usernames were selected from verified accounts of presidential candidates, former

[1] https://getdaytrends.com/peru/5/.

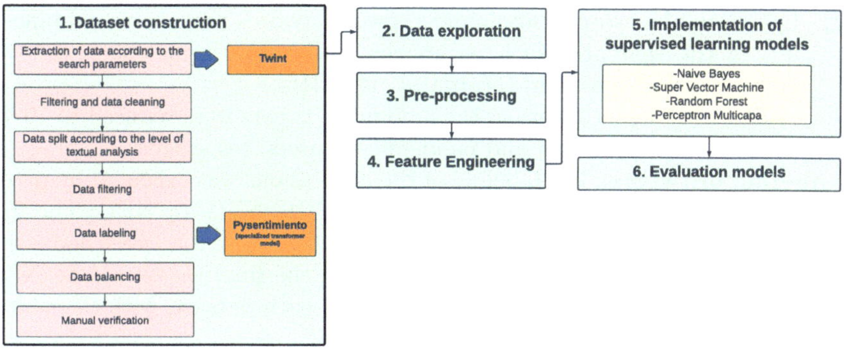

Fig. 1. Proposed methodology

presidents, state institutions, former ministers of health, and opinion experts among others.

2. **Data filtering and cleaning:** Duplicate tweets were removed using cosine similarity and tweets from other countries were manually removed. It was decided that the political dataset would only contain tweets retrieved with usernames, while the pandemic dataset would contain tweets retrieved with usernames and hashtags. Thus, the contrast between the type of information was made: the political dataset contained opinions from verified accounts (such as political figures) that would contribute to a better semantic analysis of the text, while the pandemic dataset includes tweets from any person, with colloquial expressions and textual errors. As a result, 1,953 tweets were obtained for the political dataset, and 2,151 tweets for the pandemic dataset, all extracted using hashtags; whilst 7,346 tweets for the political dataset and 10,525 tweets for the pandemic dataset were extracted using usernames.
3. **Breakdown of tweets:** Some tweets were very long and contained more than four sentences, so it was chosen to break them down into sentences using the *sent tokenize* (included in *NLTK.tokenize*[2]) function. For example: the tweet "*Yo le propuse un debate urgente, como parte de la agenda popular y usted aceptó. No mienta. El pueblo es mi prioridad de vida y nada se interpondrái a escucharlo en las calles. Nos vemos el sábado 1 de Mayo, Dia Internacional del Trabajo. El pueblo es mi prioridad de vida y nada se interpondrá a escucharlo en las calles.*", was broken down into five expressions.
4. **Filtering of sentences:** Subsequent to the sentence breakdown, "loose" elements were observed, such as links and emoticons that could generate noise. Therefore, sentences with the features listed below were eliminated:
   I. Less than four words, and that start with 'https' or '(', because the breakdown generates sentences that contain only links or expressions without further information in parentheses [4].

[2] https://www.nltk.org/api/nltk.tokenize.html.

II. Less than nine words containing stopwords (words that do not add meaning to the sentence, such as articles and prepositions) such as 'vía', 'abrazo', 'día', 'sra', and (postdata) 'PD', [16].

As a result of sentence filtering, the amount of tweets were reduced to 10,424 and 15,226 for the political and pandemic datasets, respectively.

5. **Labeling of tweets:** The labeling of the expressions was executed by pysentimiento according to a score of polarity in POS, NEG, or NEU. The final tag was the highest score in each category, for example: if a tweet had scores of POS = 0.43, NEG = 0.71, and NEU = 0.25, the final tag would be NEG. Table 1 shows the distribution of categories for each dataset, with the neutral category predominating.

Table 1. Summary of the distribution of sentences by dataset

	Political	Pandemic
Positive (POS)	1509	1109
Negative (NEG)	3558	3739
Neutral (NEU)	5357	10378
Total	10424	15226

6. **Balancing of classes:** Taking into account that the amount of data per category was not proportional (Table 1), certain criteria were applied to filter the most relevant sentences and balance the categories. Sentences that had zero likes- due to the fact that the likes are an indicator of relevance in Twitter-as well as those with less than four words were deleted. For the political dataset, only sentences that had a score greater than 0.90 in the neutral and negative categories were selected. For the pandemic dataset, only sentences with a score greater than 0.95 in the neutral and negative categories were selected. As result, 6,453 sentences for the political dataset and 4,342 sentences for the pandemic dataset were obtained.
7. **Manual verification:** A manual verification was performed on 801 tweets of the political dataset and on 906 of the pandemic dataset, which correspond to those tweets whose scores did not exceed 0.75. Those tweets were made available to three individuals-referred to as "labelers"-to conduct the labeling review, over two hours for 31 days, allocating approximately two minutes for the review of each tweet. Each labeler assigns a target and relevance to each tweet. The target feeling was defined between POS, NEG, or NEU, while relevance is binary and indicates whether it is irrelevant (1) or not (0). Sentences that did not make sense or did not correspond to the context in question were considered irrelevant. The majority vote of the three labelers

was considered as the final target and the final relevance. If there was a tie, the tweet was discarded for not having unanimous sentiment. As a final result, two datasets were obtained: political and pandemic, both available at Kaggle[3].

3.2 Data Exploration

It was observed that there were elements such as ‘peru’ and ‘peru.’-which should have been considered as one- however, the presence of the character ‘.’ causes the second word to be considered as different. In this stage, this and other problems inherent to the analysis of the raw text were identified. For the political dataset, a vocabulary of 13,236 words was determined, of which 634 were stopwords; while for the pandemic dataset, a vocabulary of 12,543 words was determined, of which 397 were stopwords. Figures 2 and 3 show the most frequently used words in each context.

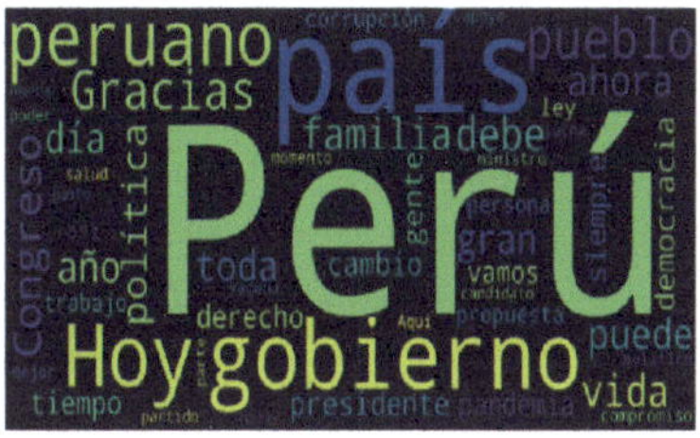

Fig. 2. WordCloud political dataset

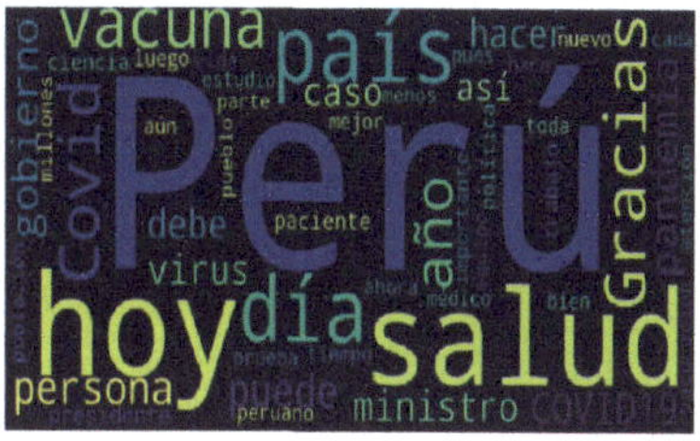

Fig. 3. WordCloud pandemic dataset

3.3 Pre-processing

At this stage, the data were pre-processed to reduce inconsistencies in the text. Therefore, the treatment of special elements (hashtags, usernames, url’s, and punctuation signs), repetition, and punctuation signs was carried out, and stemming was applied.

1. **Treatment of special elements:** Table 2 shows the labels with which the special elements such as hashtags, usernames, URL’s, and punctuation marks have been replaced. For example, for the expression `‘‘@KeikoFujimori nunca seras presidenta’’`, the following result was obtained: `‘‘__HASH_KeikoFujimori nunca seras presidenta’’`.
2. **Treatment of repetitions and punctuation marks:** Similarly, the punctuation marks were replaced with the labels of Table 2. For example, in the expression “estamos todos en peligro!” the following result was obtained: we

[3] https://www.kaggle.com/datasets/claudiasuarez/dataset-etiquetado-con-pysentimiento.

Table 2. Table of replaced characters

Element	Label
#	__HASH_
@	__user_
https://url	__URL_
?	__PUNC_QUES
!	__PUNC_EXCL
...	__PUNC_ELLP

are all in danger __PUNC_EXCL. The correction of repeated letters was also included. For the expression "leeeent", for example, the word "lento" was obtained, which is the correct expression.

3. ***Stemming:*** Through stemming, the roots of the words were identified and kept. In order to do this, the SnowballStemmer[4] algorithm, available for Python was used, which allows working with text in Spanish. For this process, the following steps were followed:
   I. The lower function was used to convert text that does not have a label (hashtag, username, among other elements) in lowercase letters.
   II. Words with a length of three or fewer characters were removed to avoid considering stopwords. In order to identify negations, some stopwords were kept such as: 'no', 'ni', or 'si'.
   III. The stem function was applied to get the roots of each word.

   By stemming the roots of the words were obtained. For example, 'acudamos' or 'acudiremos' are variations of the 'acudir' verb, in which the root is 'acud'.

3.4 Feature Engineering

This stage seeks to improve the analysis of the text by means of new attributes that allow more information to be extracted. It includes the creation of a dictionary of emojis, the text is then structured by applying N-grams, and finally negation treatment is performed.

1. **Creation of the emoji dictionary:** The emojis and emoticons were considered by constructing a dictionary as shown in Table 3. For example, the expression with emoticons "cuando llegaran las vacunas?" would be "cuando llegaran las vacunas ['U+2639']."
2. **Application of N-grams:** N-grams were used to structure the sequence of the words in each expression. The uni_grams, bi_grams, and tri_grams were considered as shown in Table 4.

[4] https://pypi.org/project/snowballstemmer/.

Table 3. Codes by feeling

Feeling	Code
Felicidad	U+1F601
Risa	U+1F923
Amor	U+2764
Guiño	U+1F609
U+2639	U+1F622

Table 4. Example of resulting N-grams

Entry	('Muy buen artículo.', 'POS')
Output	({'has(muy)': 1, 'has(buen)': 1, 'has(articul)': 1, 'has(muy,buen)': 1, 'has(buen,articul)': 1, 'has(muy,buen,articul)': 1, 'POS'})

3. **Treatment of negative sentences:** The words 'nunca', 'no', 'nada', 'ningũn', 'ninguno', 'ninguna' and 'tampoco' are identified. Each of them is labeled with 1 to identify the negation. Subsequent words are labeled with values less than 1 depending on their distance from the word identified as negation. This path is made from right to left and vice versa for each expression.

4 Experiments

This section presents the evaluation of traditional supervised learning models, which are used in classification in different contexts, making use of the political and pandemic datasets (Sect. 3.1) that were implemented in the present study. Four metrics that are usually used to evaluate text classification models were considered: (1) **accuracy**, to measure the performance of a classifier; (2) **precision**, to measure the purity of the accuracy in the exclusion of non-relevant elements; (3) **exhaustiveness**, to obtain the number of relevant elements recovered; and (4) **F1-score**, to obtain the harmonic average that combines precision and exhaustiveness.

The datasets were divided into three groups: 60% for training, 10% for validation, and 30% for testing. Supervised learning models NB, SVM, RF, and MP are instantiated by means of the *NLTK.classify* library.

4.1 Description of Experiments

Figure 4 shows the experiments performed based on questions with the aim of optimizing the results. Each experiment was evaluated by means of the accuracy (Exac.), precision (Prec.), exhaustiveness (Exhausus.), and F1-score. For the validation, only the accuracy was considered; and for the testing, all the metrics were considered. The project code is located in the *Github*[5] repository.

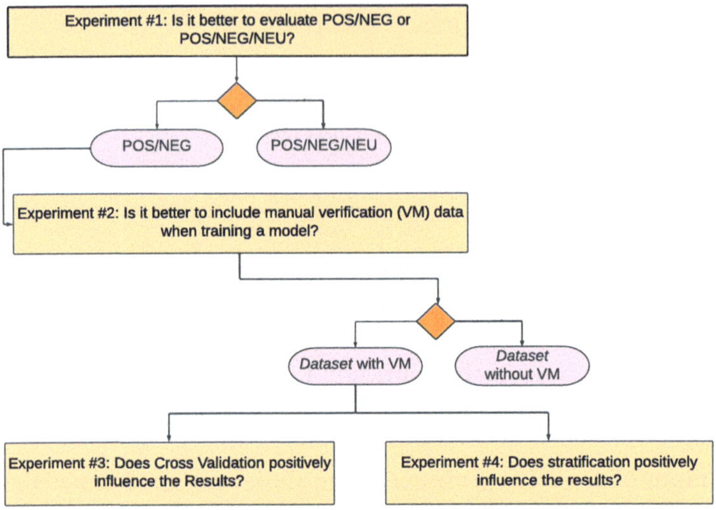

Fig. 4. Workflow for the experiments phase

Experiment #1: Is It Better to Evaluate POS/NEG or POS/NEG/NEU? A comparison between analyzing tweets as POS/NEG and POS/NEG/NEU was made to observe if the models could correctly classify in the three categories.

For the **political** dataset, Tables 5 and 6 show that the POS/NEG/NEU categories obtained lower results than POS/NEG, which indicates that the models were unable to correctly classify neutral tweets.

Table 5. Results for the political dataset based on POS/NEG classes

	Validation	Test			
	Accuracy	Accuracy	Precision	Exhaustiveness	F1-score
NB	0.782	0.791	0.686	**0.864**	**0.765**
SVM	0.822	0.766	0.750	0.608	0.672
RF	0.777	0.749	**0.843**	0.445	0.583
PM	**0.830**	**0.805**	0.830	0.635	0.719

[5] https://github.com/ClaudiaSM-12/Extraction_with_Twint.

Table 6. Results for the political dataset for POS/NEG/NEU classes

	Validation	Test			
	Accuracy	Accuracy	Precision	Exhaustiveness	F1-score
NB	0.585	0.404	**0.878**	0.404	0.533
SVM	0.588	0.732	0.426	0.632	0.438
RF	0.598	0.440	0.397	0.582	0.312
PM	**0.628**	**0.734**	0.440	**0.689**	**0.459**

For the **pandemic** dataset, Tables 7 and 8 show that POS/NEG/NEU obtains lower results than those obtained with POS/NEG, indicating again that neutral tweets were not correctly labeled.

Table 7. Results for the pandemic dataset for POS/NEG classes

	Validation	Test			
	Accuracy	Accuracy	Precision	Exhaustiveness	F1-score
NB	0.845	**0.789**	0.746	**0.795**	**0.770**
SVM	0.784	0.777	0.788	0.671	0.731
RF	0.749	0.717	**0.888**	0.415	0.565
PM	**0.868**	0.769	0.813	0.681	0.741

Table 8. Results for the pandemic dataset for the POS/NEG/NEU classes

	Validation	Test			
	Accuracy	Accuracy	Precision	Exhaustiveness	F1-score
NB	0.653	0.658	0.661	0.658	0.655
SVM	0.644	0.634	0.674	0.634	0.642
RF	0.528	0.529	0.668	0.529	0.532
PM	**0.657**	**0.682**	**0.695**	**0.682**	**0.686**

In summary, the best results were obtained when considering only the POS/NEG categories with NB and MP. It was also noticed that neutral tweets were not correctly labeled by pysentimiento. Therefore, for the following experiments only positive and negative tweets were considered.

Experiment #2: Is It Better to Include Manual Data Verification in the Training of a Model? The datasets with manually verified tweets were analyzed to determine whether verification increases the performance of each model. The results in both datasets (Tables 9 and 10) are superior to the results obtained in the previous experiment. Furthermore, NB and PM obtained optimal results for the classification of tweets in POS/NEG.

Table 9. Results of the political dataset with manual verification considering two classes

	Validation	Test			
	Accuracy	Accuracy	Precision	Exhaustiveness	F1-score
NB	0.841	0.809	0.771	**0.872**	**0.818**
SVM	0.859	0.773	0.833	0.673	0.745
RF	0.813	0.734	**0.874**	0.539	0.667
PM	**0.878**	**0.815**	0.849	0.759	0.802

Table 10. Results of the pandemic dataset with manual verification considering two

	Validation	Test			
	Accuracy	Accuracy	Precision	Exhaustiveness	F1-score
NB	**0.860**	**0.800**	0.791	**0.839**	**0.815**
SVM	0.799	0.798	0.861	0.732	0.791
RF	0.764	0.735	**0.942**	0.526	0.675
PM	0.831	0.789	0.911	0.663	0.767

In summary, including manually verified data seems to correct certain errors in the labels given by pysentimiento. Despite this, pysentimiento managed to obtain the correct label for most of the tweets, but like any model, it has a percentage of error.

Experiment #3: Does Cross Validation Positively Influence the Results? The cross validation technique was included to evaluate different supervised learning models by training different subsets of train and test, to ensure a better learning process. The results are compared with the ones in Tables 9 and 10. For both datasets, the results (Tables 11 and 12) are superior in relation to the previous experiment. Additionally, NB and MP are the models that achieved the best results.

In summary, it is observed that cross validation optimizes the results due to the execution of various models alternating different sets of train and test, which enriches the learning process.

Experiment #4: Does Stratification Positively Influence the Results? For this experiment the stratify parameter of the train test split function was

Table 11. Results for 20-folds with the political dataset

	Validation	Test			
	Accuracy	Accuracy	Precision	Exhaustiveness	F1-score
NB	0.781	**0.870**	0.763	**0.882**	**0.813**
SVM	0.848	0.842	0.772	0.754	0.759
RF	0.823	0.818	**0.815**	0.625	0.704
PM	**0.873**	0.864	0.802	0.790	0.793

Table 12. Results for 20-folds with the pandemic dataset

	Validation	Test			
	Accuracy	Accuracy	Precision	Exhaustiveness	F1-score
NB	0.835	0.831	0.735	**0.823**	**0.771**
SVM	0.830	0.827	0.777	0.733	0.750
RF	0.814	0.796	**0.814**	0.586	0.674
PM	**0.867**	**0.840**	0.799	0.732	0.759

Table 13. Results for the political dataset with class stratification

	Validation	Test			
	Accuracy	Accuracy	Precision	Exhaustiveness	F1-score
NB	0.848	**0.884**	0.834	**0.898**	**0.865**
SVM	0.831	0.847	0.851	0.762	0.804
RF	0.806	0.804	0.884	0.606	0.719
PM	**0.859**	0.877	**0.905**	0.785	0.841

Table 14. Results for the pandemic dataset with class stratification

	Validation	Test			
	Accuracy	Accuracy	Precision	Exhaustiveness	F1-score
NB	**0.861**	0.818	0.760	**0.844**	0.800
SVM	0.812	**0.850**	0.861	0.778	**0.817**
RF	0.822	0.805	**0.906**	0.611	0.730
PM	0.844	0.847	0.845	0.788	0.816

configured. This experiment applied the class stratification for the POS/NEG classes including manual verification data. It is observed that the results (Tables 13 and 14) improve compared to the previous experiment, confirming the importance of having balanced amounts of data in each category, allowing the models to learn more effectively and reducing the possibility of overfitting.

5 Conclusion

With the development of the present study, two datasets (political and pandemic) were obtained in Spanish through automatic labeling, carried out by pysentimiento, in the positive, negative, or neutral categories. In order to achieve this, a methodology for the construction of the dataset was proposed, which includes the extraction of raw tweets using Twint, data filtering and cleaning, automatic tweet labeling using pysentimiento, class balancing, and finally the manual verification of tweets. This process resulted in a consistent dataset for sentiment analysis. The experimental phase tests the effectiveness of automatic labeling only in POS/NEG categories. Likewise, the manually verified data, cross-validation, and class stratification contributed to the learning process of each model.

The present study contributes to understanding the efficiency of pysentimiento in data labeling, helping to speed up this phase in order to obtain quality data that optimizes sentiment analysis. In future studies, certain metrics could be used to assess the purity of the data. Similarly, a comparison with other labeling tools (TextBlob [13], VADER [10], etc.) could be made to provide a broader perspective and strengthen the validity of the results, as well as to delve deeper into how and why supervised learning models respond differently to pysentimiento labeling. Finally, although the methodology is focused on a specific context (Peru, political, and health events), it can be extended to other domains, regions, or languages.

Acknowledgements. This research work was carried out within the framework of the research project "Aplicación de la Ingeniería de Requisitos a los Proyectos de Investigación de la UNSAAC en desarrollo de Software y uso de TIC's", CONCYTEC - Fondecyt UNSAAC Contract No. 021-2018-UNSAAC.

References

1. Arias, F., Guerra-Adames, A., Zambrano, M., Quintero-Guerra, E., Tejedor-Flores, N.: Analyzing Spanish-language public sentiment in the context of a pandemic and social unrest: the panama case. Int. J. Environ. Res. Public Health **19**(16) (2022). https://doi.org/10.3390/ijerph191610328. https://www.mdpi.com/1660-4601/19/16/10328
2. Bekker, J., Davis, J.: Learning from positive and unlabeled data: a survey. Mach. Learn. **109**(4), 719–760 (2020). https://doi.org/10.1007/s10994-020-05877-5
3. Cañete, J., Chaperon, G., Fuentes, R., Ho, J.H., Kang, H., Pérez, J.: Spanish pre-trained bert model and evaluation data (2023). https://arxiv.org/abs/2308.02976
4. Cerón-Guzmán, J.A., León-Guzmán, E.: A sentiment analysis system of Spanish tweets and its application in Colombia 2014 presidential election. In: 2016 IEEE International Conferences on Big Data and Cloud Computing (BDCloud), Social Computing and Networking (SocialCom), Sustainable Computing and Communications (SustainCom) (BDCloud-SocialCom-SustainCom), pp. 250–257 (2016). https://doi.org/10.1109/BDCloud-SocialCom-SustainCom.2016.47

5. Chamby-Diaz, J.C., Bazzan, A.: Identifying traffic event types from twitter by multi-label classification. In: 2019 8th Brazilian Conference on Intelligent Systems (BRACIS), pp. 806–811 (2019). https://doi.org/10.1109/BRACIS.2019.00144
6. Devlin, J., Chang, M.W., Lee, K., Toutanova, K.: Bert: pre-training of deep bidirectional transformers for language understanding. In: Proceedings of the 2019 Conference of the North American Chapter of the Association for Computational Linguistics: Human Language Technologies, Volume 1 (Long and Short Papers), pp. 4171–4186 (2019). https://aclanthology.org/N19-1423.pdf
7. Giachanou, A., Crestani, F.: Like it or not: a survey of twitter sentiment analysis methods. ACM Comput. Surv. **49**(2) (2016). https://doi.org/10.1145/2938640
8. Gomez-Adorno, H., Bel-Enguix, G., Sierra, G., Barajas, J.C., Álvarez, W.: Machine learning and deep learning sentiment analysis models: case study on the sent-covid corpus of tweets in Mexican Spanish. Informatics **11**(2) (2024). https://doi.org/10.3390/informatics11020024. https://www.mdpi.com/2227-9709/11/2/24
9. Greenwade, G.D.: The comprehensive tex archive network (CTAN). TUGBoat **14**(3), 342–351 (1993). https://www.tug.org/TUGboat/tb14-3/tb40green.pdf
10. Hutto, C., Gilbert, E.: Vader: a parsimonious rule-based model for sentiment analysis of social media text. In: Proceedings of the International AAAI Conference on Web and Social Media, vol. 8, no. 1, pp. 216–225 (2014). https://doi.org/10.1609/icwsm.v8i1.14550. https://ojs.aaai.org/index.php/ICWSM/article/view/14550
11. Jimenez, R.J.S., Medishetty, B., Gao, B.J.: Compa: a comparative retrieval and analytical engine for consumer products. In: 2022 IEEE International Conference on Big Data (Big Data), pp. 6678–6681 (2022). https://doi.org/10.1109/BigData55660.2022.10020449
12. Liu, B.: Sentiment Analysis and Opinion Mining. Synthesis Lectures on Human Language Technologies, vol. 1, 1st edn. Springer, Cham (2012). https://doi.org/10.1007/978-3-031-02145-9
13. Loria, S., et al.: textblob documentation. Release 0.15 **2**(8), 269 (2018)
14. Miranda, C.H., Sanchez-Torres, G., Salcedo, D.: Exploring the evolution of sentiment in Spanish pandemic tweets: a data analysis based on a fine-tuned bert architecture. Data **8**(6) (2023). https://doi.org/10.3390/data8060096. https://www.mdpi.com/2306-5729/8/6/96
15. Muller, M., et al.: How data science workers work with data: discovery, capture, curation, design, creation. In: Proceedings of the 2019 CHI Conference on Human Factors in Computing Systems, CHI 2019, pp. 1–15. Association for Computing Machinery, New York (2019). https://doi.org/10.1145/3290605.3300356
16. Navas-Loro, M., Rodríguez-Doncel, V.: Spanish corpora for sentiment analysis: a survey. Lang. Resour. Eval. **54**(2), 303–340 (2020). https://doi.org/10.1007/s10579-019-09470-8
17. Prathap, G., Kaya, E., Palmieri, L., Korb, A., Karahan, S., Kavak, H.: A system to study anti-American misinformation and disinformation efforts on social media. In: 2022 Systems and Information Engineering Design Symposium (SIEDS), pp. 31–36 (2022). https://doi.org/10.1109/SIEDS55548.2022.9799334
18. Pérez, J.M., Furman, D.A., Alonso Alemany, L., Luque, F.M.: Robertuito: a pre-trained language model for social media text in Spanish. In: Proceedings of the Thirteenth Language Resources and Evaluation Conference (LREC 2022), Marseille, France, pp. 7235–7243. European Language Resources Association (2022). https://aclanthology.org/2022.lrec-1.785
19. Pérez, J.M., et al.: pysentimiento: a python toolkit for opinion mining and social NLP tasks (2024). https://arxiv.org/abs/2106.09462

20. Pérez, J.M., Recart Zapata, E., Alves Salgueiro, T., Furman, D., Fernández Larrosa, P.N.: A Spanish dataset for targeted sentiment analysis of political headlines. ADIO Electron. J. Inform. Oper. Res. (EJS) **22**(1) (2023). https://doi.org/10.24215/15146774e004
21. Qu, S., Yang, Y., Que, Q.: Emotion classification for Spanish with XLM-roberta and textcnn. In: IberLEF@ SEPLN, pp. 94–100 (2021). https://ceur-ws.org/Vol-2943/emoeval_paper10.pdf
22. Rateb, M.N., Alansary, S., Elzouka, M.K., et al.: Sentiment analysis in times of crisis: evaluating SVM, CNN-LSTM, and psyntimiento. arXiv preprint arXiv:2402.XXXX (v1) (2024). https://doi.org/10.21203/rs.3.rs-3994513/v1. PREPRINT (Version 1)
23. De la Rosa, J., Ponferrada, E.G., Romero, M., Villegas, P., González de Prado Salas, P., Grandury, M.: Bertin: efficient pre-training of a Spanish language model using perplexity sampling. Procesamiento del Lenguaje Natural **68**(0), 13–23 (2022). http://journal.sepln.org/sepln/ojs/ojs/index.php/pln/article/view/6403
24. Sahoo, A., Chanda, R., Das, N., Sadhukhan, B.: Comparative analysis of bert models for sentiment analysis on twitter data. In: 2023 9th International Conference on Smart Computing and Communications (ICSCC), pp. 658–663 (2023). https://doi.org/10.1109/ICSCC59169.2023.10335061
25. Tejani, A.S., Ng, Y.S., Xi, Y., Fielding, J.R., Browning, T.G., Rayan, J.C.: Performance of multiple pretrained bert models to automate and accelerate data annotation for large datasets. Radiol. Artif. Intell. **4**(4), e220007 (2022). https://doi.org/10.1148/ryai.220007

LQR Control Tuning Applied in a 2-DoF Helicopter Based on Optimization Algorithms

García Jaime Jhovani(✉), García Mejía Juan Fernando, Granda Gutiérrez Everardo Efrén, and Flores Fuentes Allan Antonio

Centro Universitario UAEM Atlacomulco, Universidad Autónoma del Estado de México, Km. 60 Carretera Toluca-Atlacomulco, 50450 Atlacomulco, Mexico
{jgarciaj,fgarciam,eegrandag,aafloresf}@uaemex.mx

Abstract. In the field of dynamic system control, precise tuning of controllers is essential for achieving optimal performance. This paper explores the tuning of the linear quadratic regulator (LQR) controller in a two-degree-of-freedom (2-DoF) helicopter using optimization techniques. The 2-DoF helicopter, known for its application in control and flight dynamics research, is employed here to evaluate the effectiveness of two optimization techniques. A genetic algorithm (GA) and particle swarm optimization (PSO) were implemented and compared to adjust the LQR controller parameters of the helicopter, aiming to enhance its performance in terms of stability and accuracy. The results show that the applied optimization methods enable effective tuning of the LQR controller, significantly improving the system's performance.

Keywords: Optimization Algorithms · Linear Quadratic Regulator · 2-DoF Helicopter

1 Introduction

A 2 degrees of freedom (2-DoF) helicopter is a scale model of a real helicopter, mounted on a fixed base with two propellers driven by two motors. The front propeller of the platform controls the nose elevation around the axis known as pitch, and the rear propeller controls the rotational movements of the helicopter around the axis known as yaw. These types of platforms are commonly used in the study of aircraft control concepts and principles, and flight dynamics [1]. In the specialized literature, numerous studies have documented the design of controllers for these helicopters, using the Quanser 2-DoF helicopter (see Fig. 1) as a testbed to analyze the effectiveness of various control strategies. These strategies include robust control [2–4], intelligent control [5,6], and optimal control [1,7,8]. The latter, due to its simplicity and high precision, has increasingly gained attention in the design of controllers based on the linear quadratic regulator (LQR).

The tuning of the LQR controller parameters, specifically the appropriate selection of the Q and R matrices, is crucial for the performance of the controlled

F. Suni-Lopez et al. (Eds.): CIIS 2024, ISEM 48, pp. 121–133, 2025.
https://doi.org/10.1007/978-3-031-91481-2_8

Fig. 1. Quanser 2-DoF helicopter model [9]

system. According to [10–14], optimization algorithms, specifically genetic algorithms (GAs) and particle swarm optimization (PSO), offer an efficient alternative for adjusting these parameters by exploring a wide range of possible solutions and converging towards one which is optimal. In this regard, the objective of this research can be defined as the comparison of a GA and a PSO for tuning the parameters of an LQR applied to the position control of a 2-DoF helicopter.

2 Modeling of 2-DoF Helicopter

The Quanser 2-DoF helicopter model is a platform anchored to a fixed vertical axis, making it a multiple-input multiple-output (MIMO) system. This system is characterized by having two points of rotation: the pitch axis (θ), and the yaw axis (ψ). By varying the voltage of two direct current (DC) motors installed at the ends of the platform, thrust forces (F_p and F_y) are generated, allowing manipulation of the helicopter's elevation movement and its rotation around the vertical axis (see Fig. 2) [15].

The dynamic behavior of the helicopter, with $p = 2$ inputs, $q = 2$ outputs and $n = 4$ state variables, is mathematically described by the equations of motion 1 and 2 [9].

$$\dot{\boldsymbol{x}}(t) = A\boldsymbol{x}(t) + B\boldsymbol{u}(t) \tag{1}$$

$$\boldsymbol{y}(t) = C\boldsymbol{x}(t) + D\boldsymbol{u}(t) \tag{2}$$

where, $\boldsymbol{x} = [\theta(t), \psi(t), \dot{\theta}(t), \dot{\psi}(t)] \in \mathbb{R}^n$ represents the state vector, $\boldsymbol{y} = [\theta(t), \psi(t)] \in \mathbb{R}^q$ is the output vector, and $\boldsymbol{u} = [u_p, u_y] \in \mathbb{R}^p$ is the control input vector corresponding to the voltages applied to the two motors [15].

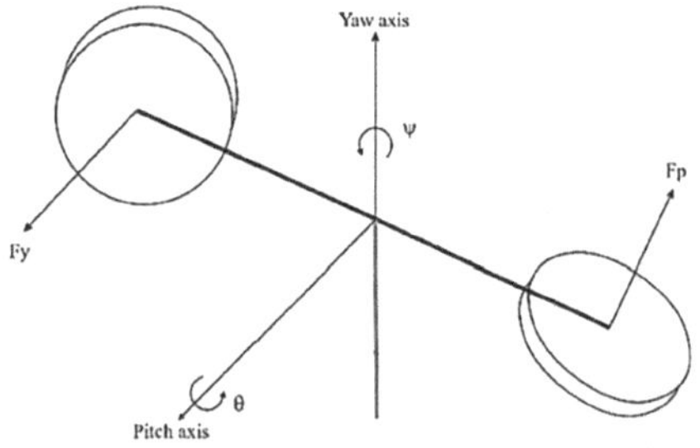

Fig. 2. Free-body diagram of the 2-DoF helicopter [15]

$A \in \mathbb{R}^{n\times n}, B \in \mathbb{R}^{n\times p}, C \in \mathbb{R}^{q\times n}, D \in \mathbb{R}^{q\times p}$ are the state matrices [9], described as follows:

$$A = \begin{bmatrix} 0 & 0 & 1 & 0 \\ 0 & 0 & 0 & 1 \\ 0 & 0 & -9.2768 & 0 \\ 0 & 0 & 0 & -0.34931 \end{bmatrix}, \quad B = \begin{bmatrix} 0 & 0 \\ 2.3667 & 0.0789 \\ 0.2406 & 0.7909 \end{bmatrix}$$

$$C = \begin{bmatrix} 1 & 0 & 0 & 0 \\ 0 & 1 & 0 & 0 \end{bmatrix}, \quad D = \begin{bmatrix} 0 & 0 \\ 0 & 0 \end{bmatrix}$$

This allows modeling of the system's motion dynamics using the scheme in Fig. 3.

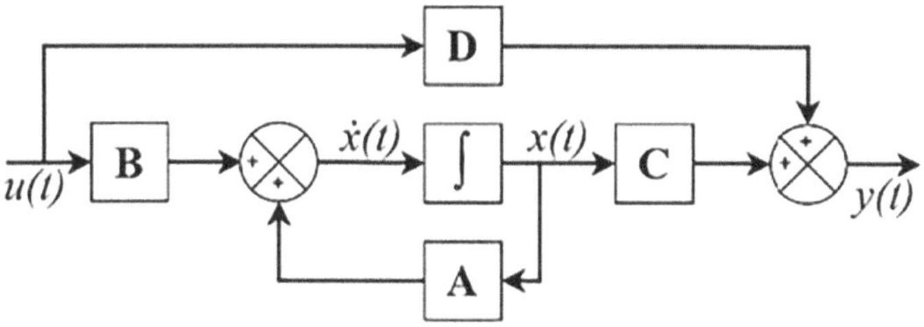

Fig. 3. State-space block diagram

3 Linear Quadratic Regulator

An LQR is an optimal control technique that minimizes a quadratic performance criterion, defined by the cost function in Eq. 3 [16].

$$J = \int_0^\infty \left(\boldsymbol{x}^T Q \boldsymbol{x} + \boldsymbol{u}^T R \boldsymbol{u}\right) dt \tag{3}$$

where Q is a symmetric positive semidefinite matrix that penalizes deviations in the system states, and R is a symmetric positive definite matrix that penalizes the magnitude of the control inputs.

To minimize the cost function J, it is necessary to find the gain matrix K, which is calculated using Eq. 4 [16].

$$K = R^{-1}B^{T}P \tag{4}$$

where P is a symmetric positive definite matrix that satisfies the Riccati algebraic equation defined as $A^{T}P + PA - PBR^{-1}B^{T}P + Q = 0$. The block diagram of the helicopter controlled by LQR is shown in Fig. 4.

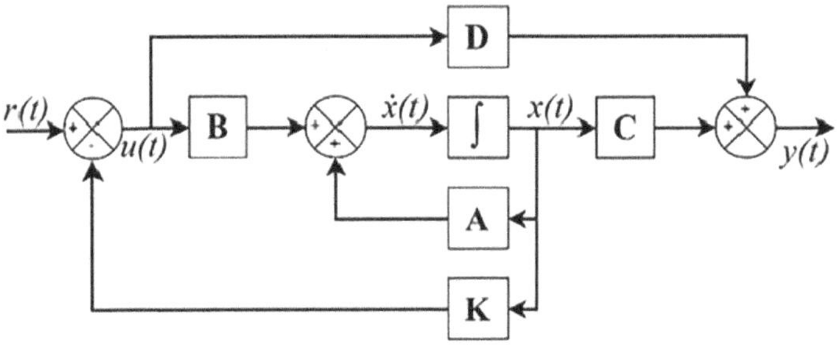

Fig. 4. Block diagram of the LQR controller

Before obtaining the optimal value of K, the weighting matrices R and Q need to be adjusted. The choice of these matrices is crucial for achieving the desired performance of the helicopter. The form of these matrices is defined as follows:

$$Q = \begin{bmatrix} q_{11} & 0 & 0 & 0 \\ 0 & q_{22} & 0 & 0 \\ 0 & 0 & q_{33} & 0 \\ 0 & 0 & 0 & q_{44} \end{bmatrix}, \quad R = \begin{bmatrix} r_{11} \end{bmatrix}.$$

The determination of the values q_{11}, q_{22}, q_{33}, q_{44}, and r_{11} can be approached classically by selecting values based on experience and knowledge of the system, and then adjusting them through trial and error. Alternatively, it can be done analytically by calculating the matrix Q as the product of C and its transpose C^{T}, and defining the matrix R as an identity matrix, such that:

$$Q = C \cdot C^{T} = \begin{bmatrix} 1\,0\,0\,0 \\ 0\,1\,0\,0 \\ 0\,0\,0\,0 \\ 0\,0\,0\,0 \end{bmatrix}, \quad R = \begin{bmatrix} 1 \end{bmatrix}$$

An alternative to this is to use optimization techniques to explore the solution space and find the most appropriate values for Q and R.

4 Optimization Algorithms

4.1 Genetic Algorithm

Genetic algorithms are an optimization and search technique inspired by the process of natural evolution. They are used to solve optimization problems where the goal is to find the best possible solution among a set of potential solutions [17]. The underlying theory of GAs is based on principles of natural selection, genetic recombination, and mutation as described by Darwin and Mendel, which stipulate the conditions for the survival of the fittest individual in an environment. The idea of using these principles was proposed by the American computer scientist John Holland in the 1960s [18].

Holland proposes finding the solution to an optimization problem starting from a collection of possible solutions $P = [C_1, C_2, \ldots, C_n]$, known as the population. Each individual in P, referred to as a chromosome, is composed of genes that form the vector $C_i = [g_1, g_2, \ldots, g_n]$. These genes encode the characteristics of each solution [19]. The individuals in this population undergo a selection process, based on evaluating the fitness of everyone with respect to the specific objective function of the problem, and are then subjected to operations of crossover, mutation, and replacement.

These operations are performed iteratively until a predefined stopping criterion (number of generations) is reached. In the final generation, the chromosome that produces the highest fitness value is considered the solution to the optimization problem [20]. The steps characterizing a GA are shown in Fig. 5.

4.2 Particle Swarm Optimization

Particle swarm optimization is an optimization technique developed by James Kennedy and Russell Eberhart in 1995, inspired by the collective behavior of swarms of animals, such as flocks of birds or schools of fish. This algorithm utilizes a population of particles that move through the search space, adjusting their positions and velocities based on personal experience and the collective experience of the swarm [21].

Each particle represents a possible solution to the optimization problem and has a position $\boldsymbol{p} = [p_1, p_2, \ldots, p_n]$ and a velocity $\boldsymbol{v} = [v_1, v_2, \ldots, v_n]$ in the search space. The particles move iteratively by adjusting their velocity and position based on two factors: the best personal position found by each particle (p_{best}) and the best global position found by any particle in the swarm (g). For each

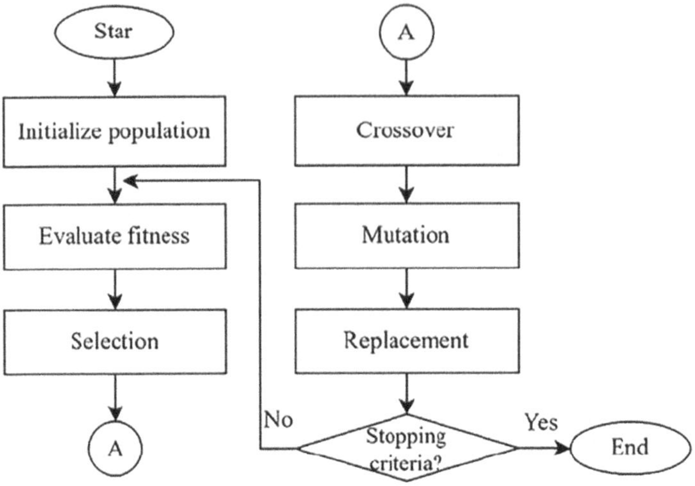

Fig. 5. Flowchart of a GA

$\boldsymbol{p}_i$, the equation that updates its velocity is defined as in Eq. 5. Subsequently, it updates its position using the expression in Eq. 6 [22].

$$v_i(t+1) = wv_i(t) + c_1r_1\left(p_{\text{best}}(t) - p_i(t)\right) + c_2r_2\left(g(t) - p_i(t)\right) \tag{5}$$

$$p_i(t+1) = p_i(t) + v_i(t+1) \tag{6}$$

where w is the inertia weight factor, c_1 and c_2 are the cognitive and social acceleration coefficients, respectively, which determine the influence of the best personal position and the best global position. r_1 and r_2 are vectors of random values between 0 and 1 that introduce variability into the process [22].

In each iteration, the particles evaluate their position using the objective function specific to the problem. Simultaneously, the swarm updates the best global position if any particle finds a solution that surpasses the best global position known so far. This iterative process continues until a predefined stopping criterion is met, such that, in the final iteration, the best global position is considered the solution to the problem [23]. The steps and dynamics of PSO are illustrated in Fig. 6.

5 Simulation and Results

With the objective of being able to simulate and obtain the results documented in this proposal, the methodological scheme shown in Fig. 7 was implemented. First, the mathematical model of the helicopter and the objective function to be

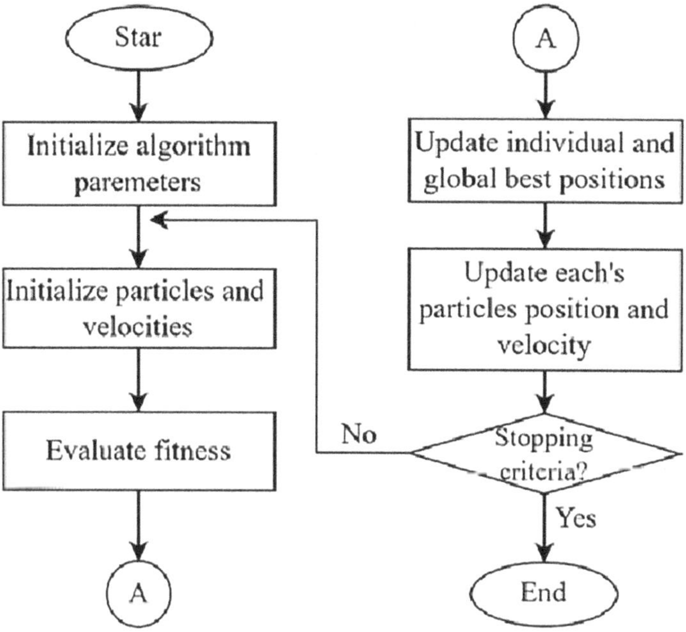

Fig. 6. Flowchart of a PSO

evaluated are obtained. With this, an LQR is designed through the analytical approach. Afterwards, PSO and GA are applied to tune the controller parameters. Finally, the controller is validated and a comparative statistical study of the results obtained is carried out.

For the design of the GA and PSO, the parameters were configured as shown in Table 1. It is important to mention that the crossover operator for the GA was the BLX-α crossover, which for each pair of parent chromosomes $C_1 = [g_{11}, g_{12}, \ldots, g_{1n}]$ and $C_2 = [g_{21}, g_{22}, \ldots, g_{2n}]$, generates two descendant chromosomes $H_1 = [g_{11}, g_{12}, \ldots, g_{1n}]$ and $H_2 = [g_{21}, g_{22}, \ldots, g_{2n}]$. Thus, each gene g is generated in the interval $[C_{\min} - I \cdot \alpha, C_{\max} + I \cdot \alpha]$, where $C_{\min} = \min(g_{1i}, g_{2i})$, $C_{\max} = \max(g_{1i}, g_{2i})$, $I = C_{\max} - C_{\min}$, and $\alpha = 0.5$. On the other hand, the uniform mutation operator is used to replace a random gene g from three randomly chosen descendant chromosomes H_i by changing its value to another determined by the interval of the search space.

The objective function of the optimization algorithms aims to minimize the sum of the integral and quadratic errors of the system. This is described by Expression 7.

$$F_{\text{obj}} = \frac{1}{(IAE + ISE + ITAE + ITSE) + 1} \tag{7}$$

where, the integral of time-weighted absolute error (ITAE), the integral of time-weighted squared error (ITSE), the integral of squared error (ISE), and the

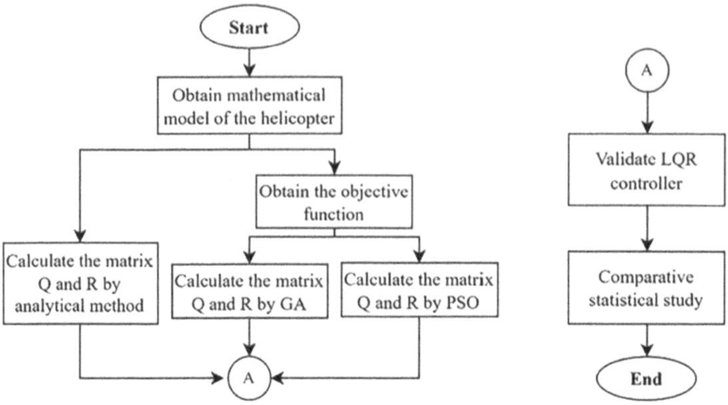

Fig. 7. Proposed methodological scheme

Table 1. Parameters of GA and PSO

Parameter	GA	PSO
Search space	[0, 10]	[0, 10]
Variables	$[q_{11}, q_{22}, q_{33}, q_{44}, r_{11}]$	$[q_{11}, q_{22}, q_{33}, q_{44}, r_{11}]$
Number of iterations	100	100
Population size	60	60
Crossover rate	80%	N/A
Mutation rate	5%	N/A
Inertia factor (w)	N/A	0.5
Cognitive coefficient (c_1)	N/A	1
Social coefficient (c_2)	N/A	2

integral of absolute error (IAE) are considered. These are described by the following Eqs. 8, 9, 10, and 11, respectively. The addition of 1 in the denominator prevents division by zero and ensures that the objective function value is always positive.

$$IAE = \int_0^T |e(t)|\, dt \tag{8}$$

$$ISE = \int_0^T e^2(t)\, dt \tag{9}$$

$$ITAE = \int_0^T t \cdot |e(t)|\, dt \tag{10}$$

$$ITSE = \int_0^T t \cdot e^2(t)\, dt \tag{11}$$

Figure 8 shows an example of the convergence curves for GA vs PSO. This pattern suggests that, although the difference in the final values of the objective function between the two algorithms is minimal (0.01%), the GA can explore the search space more rapidly compared to the PSO. These results are summarized in Table 2.

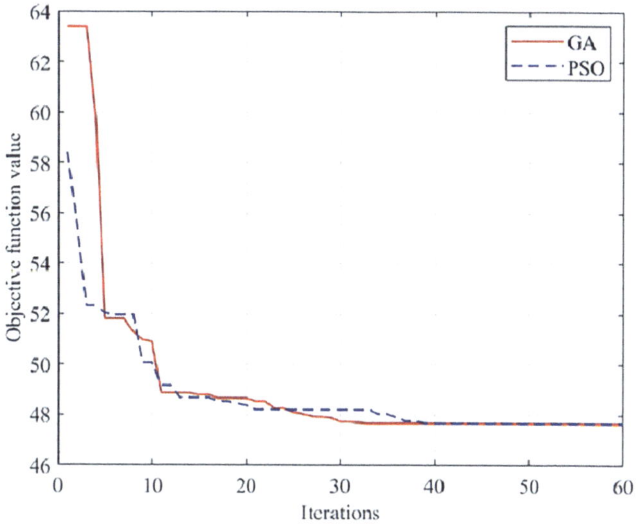

Fig. 8. Comparison of the convergence curves of GA and PSO

Table 2. Comparison of the convergence of GA and PSO

Parameter	GA	PSO
Objective function value	47.6808	47.6682
Error sum reduction %	68.78%	68.79%
Convergence iteration	33	40

Each algorithm was evaluated iteratively 40 times to statistically analyze its performance. The results (see Table 3) were subjected to Kolmogorov-Smirnov and Shapiro-Wilk normality tests, which indicated that the data do not follow a normal distribution. Subsequently, the Friedman test was applied to assess the statistical significance of the differences between the algorithms, finding that there are no statistically significant differences. Additionally, the median and interquartile range were calculated to analyze the distribution and dispersion of the data. These results show that, although the differences in medians are minimal, the interquartile range of the GA is smaller, suggesting greater consistency in its results.

Table 3. Statistical analysis

	GA	PSO
Kolmogórov-Smirnov test	1.08×10^{-36}	1.48×10^{-36}
Shapiro-Wilk test	1.15×10^{-9}	5.62×10^{-8}
Friedman test	0.1138	
Median	47.7017	47.6897
Quartile 1	47.6919	47.6701
Quartile 3	47.7388	48.0766
Interquartile range	0.0469	0.4065

Finally, the parameters that tune the matrices Q and R are the result of the median of the best chromosomes and particles from the 40 executions, which can be observed in Table 4.

Table 4. Tuned Q and R matrices

Matrix	GA	PSO
Q	$\begin{bmatrix} 3.0865 & 0 & 0 & 5.7311 \\ 0 & 0 & 0 & 0 \\ 0 & 0 & 0 & 0 \\ 0.2770 & 0 & 0 & 0.2415 \end{bmatrix}$	$\begin{bmatrix} 4.5220 & 0 & 0 & 8.3955 \\ 0 & 0 & 0 & 0 \\ 0 & 0 & 0 & 0 \\ 1.1075 & 0 & 0 & 0.9795 \end{bmatrix}$
R	$\begin{bmatrix} 4.4317 \end{bmatrix}$	$\begin{bmatrix} 6.4943 \end{bmatrix}$

The responses of the pitch angle (θ) and yaw angle (ψ) of the 2-DoF helicopter, under different control configurations, using the parameters optimized by each algorithm are shown in Fig. 9 and summarized in Table 5.

Table 5. Performance comparison

Performance criteria	LQR		LQR+GA		LQR+PSO	
	θ	ψ	θ	ψ	θ	ψ
Settling time	18 s	20 s	20 s	18 s	20 s	18 s
Steady state error	−12 %	13 %	−0.10 %	−0.55 %	−0.10 %	−0.59 %
IAE	7.3968	9.4479	3.9479	3.6034	3.9624	3.6141
ISE	2.3208	2.9256	2.0220	1.9541	2.0295	1.9593
ITAE	141.9408	173.5226	15.2644	11.7716	15.3904	11.8461
ITSE	18.0214	26.1196	3.9064	3.3272	3.9359	3.3469

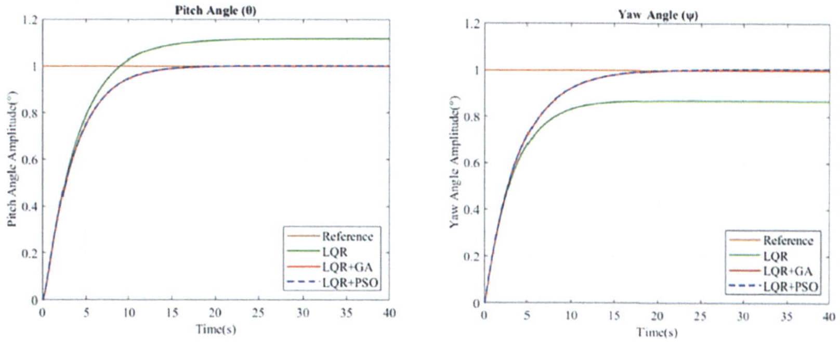

Fig. 9. Time response comparison of the pitch axis (θ) and yaw axis (ψ) of the helicopter under different control configurations: LQR, LQR+GA, and LQR+PSO.

6 Conclusion

The results obtained demonstrate that it is possible to tune the Q and R matrices of the linear quadratic regulator (LQR) used in the control of the 2-DoF helicopter through optimization techniques, specifically a genetic algorithm (GA) and particle swarm optimization (PSO). Both techniques minimized the performance criteria IAE, ISE, ITAE, ITSE as shown in Table 5. However, a greater reduction is shown for the ITAE in θ and ψ angles with percentages of 89% and 93%, respectively, compared to the analytical adjustment of the LQR controller. Analysis of the convergence curves and statistical study reveals that both GAs and PSO exhibit similar performance in terms of settling time and steady-state error. However, it was observed that the GA has lower variability in its results and converges more quickly compared to the PSO.

For future work, it is proposed to evaluate the performance of the optimization algorithms used with different configurations or recombination operators, as well as making comparisons with other optimization algorithms not considered in this study. Additionally, it would be valuable to explore the combination of the LQR controller with other control techniques to accelerate the response times of the 2-DoF helicopter.

References

1. González, E.C.V., Rivera, D.M., Gómez, E.J.: Model and observer-based controller design for a quanser helicopter with two DoF. In: 2012 IEEE Ninth Electronics, Robotics and Automotive Mechanics Conference, pp. 267–271 (2012). https://doi.org/10.1109/CERMA.2012.50
2. Khayati, K.: Multivariable adaptive sliding-mode observer-based control for mechanical systems. Can. J. Electr. Comput. Eng. **38**(3), 253–265 (2015). https://doi.org/10.1109/CJECE.2015.2406873
3. Borkar, A., Patil, P.M.: Super twisting observer based full order sliding mode control. Int. J. Dyn. Control **9**(4), 1653–1659 (2021). https://doi.org/10.1007/s40435-021-00757-9

4. Watanabe, Y., Katsurayama, N., Takami, I., Chen, G.: Robust LQ control with adaptive law for MIMO descriptor system. In: 2013 9th Asian Control Conference (ASCC), pp. 1–6 (2013). https://doi.org/10.1109/ASCC.2013.6606248
5. Emami, S.A., Castaldi, P., Banazadeh, A.: Neural network-based flight control systems: present and future. Annu. Rev. Control **53**, 97–137 (2022). https://doi.org/10.1016/j.arcontrol.2022.04.006. https://www.sciencedirect.com/science/article/pii/S1367578822000219
6. Chaudhary, A., Bhushan, B.: Trajectory tracking of a 2-DoF helicopter system using fuzzy controller approach. In: 2021 International Conference on Emerging Techniques in Computational Intelligence (ICETCI), pp. 159–164 (2021). https://doi.org/10.1109/ICETCI51973.2021.9574049
7. Karthick, S., Kanthalakshmi, S., Vinodh Kumar, E., Joshi Kumar, V., Ezhil Kumaran, A.: Experimental validation of adaptive augmented LQI control for a 2 DoF helicopter. IETE J. Res. **70**(1), 584–596 (2024). https://doi.org/10.1080/03772063.2022.2138578
8. Ganapathy Subramanian, R., Elumalai, V.K.: Robust MRAC augmented baseline LQR for tracking control of 2 DoF helicopter. Robot. Auton. Syst. **86**, 70–77 (2016).https://doi.org/10.1016/j.robot.2016.08.004. https://www.sciencedirect.com/science/article/pii/S0921889015302864
9. Quanser Inc.: Quanser 2-DoF helicopter laboratory manual (2011). https://users.encs.concordia.ca/~realtime/elec483/docs/2%20DOF%20Helicopter%20Lab%20Manual.pdf. Accessed 3 Sept 2024
10. Yu, W., Li, J., Yuan, J., Ji, X.: LQR controller design of active suspension based on genetic algorithm. In: 2021 IEEE 5th Information Technology,Networking,Electronic and Automation Control Conference (ITNEC), vol. 5, pp. 1056–1060 (2021). https://doi.org/10.1109/ITNEC52019.2021.9587272
11. Vishal, Ohri, J.: GA tuned LQR and PID controller for aircraft pitch control. In: 2014 IEEE 6th India International Conference on Power Electronics (IICPE), pp. 1–6 (2014). https://doi.org/10.1109/IICPE.2014.7115839
12. Al-Mahturi, A., Wahid, H.: Optimal tuning of linear quadratic regulator controller using a particle swarm optimization for two-rotor aerodynamical system. Int. J. Electron. Commun. Eng. **11**(2), 196–202 (2017). https://doi.org/10.5281/zenodo.1128899
13. Fang, J.: The LQR controller design of two-wheeled self-balancing robot based on the particle swarm optimization algorithm. Math. Probl. Eng. **2014**(1), 729095 (2014). https://doi.org/10.1155/2014/729095. https://onlinelibrary.wiley.com/doi/abs/10.1155/2014/729095
14. Önen, U., Çakan, A., İlhan, I.: Performance comparison of optimization algorithms in LQR controller design for a nonlinear system. Tur. J. Electr. Eng. Comput. Sci. **27**(3) (2019). https://doi.org/10.3906/elk-1808-51. https://journals.tubitak.gov.tr/elektrik/vol27/iss3/27
15. Ouyang, Y., Dong, L., Xue, L., Sun, C.: Adaptive control based on neural networks for an uncertain 2-DoF helicopter system with input deadzone and output constraints. IEEE/CAA J. Automatica Sinica **6**(3), 807–815 (2019). https://doi.org/10.1109/JAS.2019.1911495
16. Fadzilah Hashim Mohd Yusoff Mashor, S.M.S.: Full- and reduced-order compensator for innosat attitude control. Jurnal Teknologi (Sci. Eng.) **76**(12) (2015). https://doi.org/10.11113/jt.v76.5883. https://journals.utm.my/jurnalteknologi/article/view/5883

17. Kramer, O.: Genetic Algorithm Essentials, 1 edn. Studies in Computational Intelligence. Springer Cham (2017). https://doi.org/10.1007/978-3-319-52156-5. EBook ISBN: 978-3-319-52156-5 (Published: 07 January 2017), Softcover ISBN: 978-3-319-84834-1 (Published: 13 July 2018), Hardcover ISBN: 978-3-319-52155-8 (Published: 13 January 2017), Series ISSN: 1860-949X, Series E-ISSN: 1860-9503
18. Goldberg, D.E.: Genetic Algorithms in Search, Optimization and Machine Learning. Addison-Wesley Pub. Co, Massachusetts (1989)
19. Mitchell, M.: An Introduction to Genetic Algorithms, 7th edn. MIT Press, Cambridge (2001)
20. Gen, M., Cheng, R.: Genetic Algorithms and Engineering Optimization. Wiley-Interscience, New York (2000)
21. Kennedy, J., Eberhart, R.: Particle swarm optimization. In: Proceedings of ICNN'95 - International Conference on Neural Networks, vol. 4 (1995). https://doi.org/10.1109/ICNN.1995.488968
22. Poli, R., Kennedy, J., Blackwell, T.: Particle swarm optimization. Swarm Intell. **1**, 33–57 (2007). https://doi.org/10.1007/s11721-007-0002-0
23. Yousaf, S., Mughees, A., Khan, M.G., Amin, A.A., Adnan, M.: A comparative analysis of various controller techniques for optimal control of smart nano-grid using GA and PSO algorithms. IEEE Access **8**, 205696–205711 (2020). https://doi.org/10.1109/ACCESS.2020.3038021

A Comprehensive Market Price Analysis Using Python

Fernando Baca Ordónez(✉), Camila Jennifer Chong Carmen, Bruno Gerwer Campero Del Pozo, Patrick Henry Cordova Velarde, Abner Samuel Huinapi Huaman, and Victor Andres Ayma Quirita

Universidad del Pacífico, Lima, Peru
{f.bacao,cj.chongc,bg.camperod,ph.cordovav,as.huinapih}@alum.up.edu.pe, va.aymaq@up.edu.pe

Abstract. The Latin American capital market is still in development and faces challenges in infrastructure, innovation, efficiency, and financial integration. It is important to address these problems due to their impact on the economic development of countries and their relation to social well-being. Therefore, this work addresses the widespread need for access to information regarding the financial sector by creating a Python platform that facilitates people's investment decisions. The methodology is based on the recreation of financial models (such as Markowitz) and technical analysis strategies (such as VixFix, moving averages, and momentum) in Python, so users could have the opportunity to test on asset prices extracted from Yahoo Finance-including stocks, ETFs, cryptocurrencies, forex, and futures under the different models to evaluate their performance when compared to buying them passively. From this, an easy-to-manipulate console was created for processing the information using the mentioned algorithms and providing easy-to-understand graphical results. This allows the evaluation of asset performance under different strategies that could offer superior returns compared to other passive income methods. This article represents a promising proposal for automation and analysis in the stock market that could potentially serve as a free-access tool for price analysis in less liquid and shallow markets in the future.

Keywords: Financial inclusion · market · portafolio management · Python

1 Introduction

The financial inclusion gap is a persistent challenge in Latin America and the Caribbean, where only 39% of the population has an account with a formal financial institution, and just 8% have access to formal financing [23]. This disparity underscores the need for innovative solutions that democratize investment opportunities.

F. Suni-Lopez et al. (Eds.): CIIS 2024, ISEM 48, pp. 134–146, 2025.
https://doi.org/10.1007/978-3-031-91481-2_9

Authors such as Rodríguez [24] highlight the potential for bringing capital markets closer to the public, since it stimulates economic growth through liquidity. However, in practice, limited financial education and familiarity with investment strategies on existing platforms pose significant obstacles. Despite the diversity of investment instruments and strategies, the current literature lacks comprehensive studies that implement and demonstrate their use and effectiveness in an integrated way [4].

In this context, the project focuses on providing users with the ability to build portfolios tailored to their goals, addressing the following needs:

- Personalization of strategies: There are various investment platforms that offer predefined strategies. For example, eToro broker allows users to automatically copy trades from more experienced investors [5], while TradeStation enables automated trading but with limitations on indicators [21]. This project overcomes the predefined strategies restrictions so that users can obtain more precise information when considering their individual goals and preferences.
- Accessibility to a variety of financial models: Implementing diverse methods allows users to diversify their decision-making criteria to make more informed decisions, testing the performance of these strategies on selected assets.
- Interactivity and flexibility: The ability to select specific assets and the freedom to adjust parameters ensure an experience that is entirely tailored to the user and their decision-making process.
- Integration with Yahoo Finance data: Implementing Python packages eliminates the extra work of manually downloading data for the user [22].
- Accessibility for beginners: The project offers intuitive and efficient tools to optimize portfolios (factors with standard parameters and constraints), even for users who are not experienced in portfolio management.

In this sense, this project improves portfolio management platforms, making investment more accessible, informed, and personalized, thus redefining the user experience with flexible financial models. Without approaches like ours, economic gaps-such as limited financial inclusion and socioeconomic inequality-would persist as a significant portion of the population would remain outside the formal financial system, considering that limited access to information and low understanding of investment opportunities restrict economic growth due to under-utilization of resources [7]. In addition, investors may be discouraged by the inability to diversify efficiently and the high risk associated with capital accumulation, due to the inaccessibility of portfolio management platforms [25].

However, strict adherence to a single portfolio management method can lead to inefficient strategies due to modeling risk, as theoretical models rely on many restrictive assumptions [6]. By offering various financial methods, significant monetary losses could be avoided, as some models excel at maximizing profitability but may not adequately address risk, such as Markowitz (see [6] and [10]). Addressing these challenges is essential for improving the capital market. This project introduces new tools that streamline tedious implementations and optimize the visualization of results to prevent deviations from optimal choices [14]. In addition, it will improve financial education, empower investors, promote efficient use of resources, and provide a stronger knowledge base, allowing the scientific community to gain a more holistic understanding of the capital market.

The main objective of this paper is to simplify asset price analysis through programming tools, updated data, financial models, and statistical calculations, providing a comprehensive view for those interested in exploring various forms of investment using all available information. The specific objectives include linking up-to-date data that enables price analysis and comparison with historical price data retrieved from credible online sources; implementing a range of financial models based on statistical calculations and approaches proposed by experts, to effectively determine asset prices; generating visually appealing graphics and concise explanations to complement the findings and aid in evaluating whether asset prices will yield higher returns, making it easier for users without prior financial experience to understand; and identifying asset price trends across different markets to support decision-making based on historical behavior and financial theory.

This work aims to provide easy-to-use price analysis tools so that companies can channel their capital, fostering innovation and driving economic growth. It seeks to reduce information asymmetry and bridge the financial inclusion gap in the capital market, as well as improve financial literacy for informed investment decisions. By applying programming techniques and analysis of market price patterns, promising assets can be identified to yield higher returns, empowering individuals in resource management. In addition, the implementation of the aforementioned algorithms will identify opportunities and risks in the capital markets, allowing for more strategic decision-making.

2 Literature Review

Financial models have been extensively studied in academia and have been crucial for capital management for both individuals and companies [16]. These tools aim to empower users to efficiently implement financial models that align with their investment needs and goals. Below is a brief review of theoretical contributions to finance and their contemporary applications.

First, Harry Markowitz revolutionized investment theory with his influential work, "Portfolio Selection", published in the Journal of Finance. He introduced the concept of balancing risk and return, emphasizing the need to invest in the market to maximize returns with the least possible risk [15].

William F. Sharpe, John Lintner, and Jan Mossin made significant contributions to finance with their works "Capital Asset Prices: A Theory of Market Equilibrium under Conditions of Risk", "The Valuation of Risk Assets and the Selection of Risky Investments in Stock Portfolios and Capital Budgets", and "Equilibrium in a Capital Asset Market", respectively. Collectively, they characterized the Capital Asset Pricing Model (CAPM). This model provides a theoretical framework for evaluating the relationship between risk and expected return, being an essential tool for investors and financial analysts (see [13,27], and [18]).

Furthermore, Larry Williams introduced the VixFix model in "Long-Term Secrets to Short-Term Trading," an indicator designed to detect extreme changes in market trends, focusing on volatility. The model provides investors with a unique tool to make informed decisions about the timing of market entry or exit [28]. Various factors have also been studied to explain the long-term risk and return of an asset. For example, momentum is considered a persistence factor in assets and indicates that they will perform the same way as they did in the short term [17]. Another example is the case of moving averages, which is a trend following device that helps better identify the beginning and end of trends [19]. This aligns with the theories of trend stationarity and mean reversion through smoothing of the curves.

Recently, Yves Hilpisch has played a pivotal role in promoting and applying quantitative tools in finance through his work "Python for Finance: Mastering Data-Driven Finance." He provides a comprehensive guide to using Python in the realm of quantitative finance. His contribution forms a valuable foundation for this project, as it formalizes the processes from effective financial data download to the practical implementation of complex quantitative models [12].

This project represents a significant advancement by integrating four financial models into a single accessible platform. Unlike previous works, it offers users the flexibility to explore the outcomes of various models, allowing for more informed and personalized decision making. Additionally, it leverages Python's programming capabilities and the availability of real-time data to provide a dynamic tool tailored to individual needs.

3 Theoretical Framework

Latin American economies are currently in the process of developing their capital markets. At present, markets are narrow, which can lead to fragility and volatility in response to changes in the local and international economic and financial environment [2]. In this scenario, the authors argue that there is an urgent need for specific enhancement of the structure and efficiency of capital markets, as this is not inherently linked to overall economic development.

Grupo Sura highlights significant challenges faced by the region's capital markets, including the need to stimulate innovation and efficiency, as well as to better measure risk and estimate the supply of financial products. These efforts aim to build a more modern market capable of meeting investor needs. Despite economic cycles and occasional setbacks, many investors maintain an active buying position, which presents a valuable opportunity to consolidate regional economic development [11].

To facilitate understanding of the project and its implementations, it is relevant to define the following concepts:

3.1 Capital Market

A space that offers an alternative to traditional bank funding by issuing and trading financial products. In doing so, related expenses are reduced, and investment returns increase [3].

3.2 Markowitz Model

A method for portfolio optimization that seeks the best possible balance between risk reduction and expected return. It takes into account the preferences of the investors as shown by their indifference curves. The strategy involves shifting ideal asset combinations to the efficient frontier by diversifying assets with negative correlations, reducing the overall risk of the portfolio [16].

3.3 VixFix

An index that compares the current closing price with a range of historical values in order to assess market expectations of volatility, which is a reflection of the uncertainty facing investors. When the closing price is below the minimum range, it shows minimal volatility; when it is above the maximum range, it indicates significant volatility. It gives investors a critical view on the stability of the market and directs decisions about the entry and exit of positions [8].

3.4 Momentum

Allows anticipating changes in market trends based on the premise that assets with good past performance will continue to do so in the future. This involves buying assets with recent good performance and selling those with poor performance, using historical prices as the main variable for its calculation [9].

3.5 Moving Averages

An indicator that smooths the price curve to reduce abrupt variations caused by volatility. This provides a more suitable profile for following trends and generating buy and sell signals. In addition, it reflects the consensus of investors on the value of an asset at any given time. It is calculated considering closing prices for a specific period, either simply or exponentially [1].

4 Materials and Methods

To facilitate the analysis of asset prices, an interface has been created to serve as a communication medium between users and the models described in the previous sections of this work. This was done using the Python programming language and a list of five libraries. These include Yfinance, used to download databases with asset prices; numpy, implemented for performing mathematical operations; pandas, used for data manipulation; datetime, used for working with dates; and matplotlib for creating charts.

Figure 1 presents a description of the methodology used to develop the Python program. First, all the libraries that will be used throughout the Python code are imported. The first five markets available in the TradingView application are then printed in the console as a reference. These are divided into stocks, ETFs, crypto, forex, and futures. From this, the user must choose the index number of the market they would like to enter. Subsequently, the four models recreated in the work will be printed in the console: Markowitz, VixFix, momentum, and moving averages. The user chooses the model in the same way they selected the market. After selecting the model, five assets available to the user will be displayed on the screen. If the selected model is different from the Markowitz model, the user will be able to choose one of the five assets. This is because, in the Markowitz model, a portfolio is recreated where the five shown assets are compared, while the other models work with only one asset.

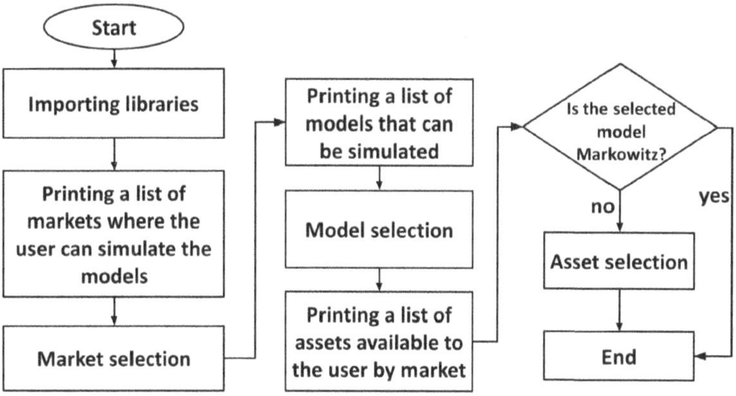

Fig. 1. Market, model, and asset selection for analysis.

The process described in the previous paragraph represents the interaction between the user and the created program. Based on all the stored information, the selected model is reconstructed. To do this, the asset database is downloaded from the Yfinance library using tickers. Additionally, the datetime package is used to download prices over a 200-day period up to the date the code is run. The models are then reconstructed based on the bibliography revolving around the sources of the state of the art, and the charts obtained from these models are printed. Depending on the value of the accumulated returns of the model, whether positive or negative, a comment is printed along with a recommendation. This process can be seen in Fig. 2.

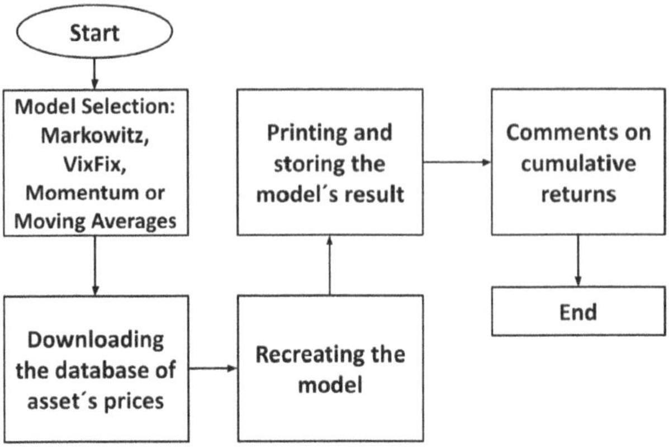

Fig. 2. Results of the returns

5 Results

After implementing the algorithms presented in the methodology, the output from the console interaction with the user will be displayed. These results consist of images showing the outcomes after testing the different recreated models. This way, users can base their investment strategies on the models that provide the most desirable results according to their preferences.

If the user selects the stock market under the Markowitz model in the console, an output graph like Fig. 3 will be displayed, where the accumulated returns of the stocks listed for the user can be visualized. This shows that among the five selected stocks, NVDA had the highest return in the last 200 days, exceeding 50% annualized return. The stock that performed the worst was TSLA along with AAPL. Based on this graph, the user can make decisions based on the set of assets that performed the best. However, this does not guarantee that positive returns will continue, so a risk analysis should also be performed considering the volatility of the assets.

If the user had chosen the cryptocurrency market instead of the stock market, a graph similar to Fig. 3 would have appeared, showing the accumulated returns from the passive purchase of the most important cryptocurrencies in Trading View. Then, a graph like Fig. 4 would have appeared, showing the results of the Markowitz model.

In this figure, the cryptocurrencies Solana and Bitcoin stand out, as the former had a return of approximately 200% annualized over 200 days, while the latter is located right on the efficient frontier, indicating that it is the asset with the highest return for the lowest risk, with a 75% return and 0.2 volatility in the last 200 days of the analyzed period. The remaining points correspond to combinations with random weights to design a portfolio with the available assets in the cryptocurrency market. The most preferable are those that correspond to

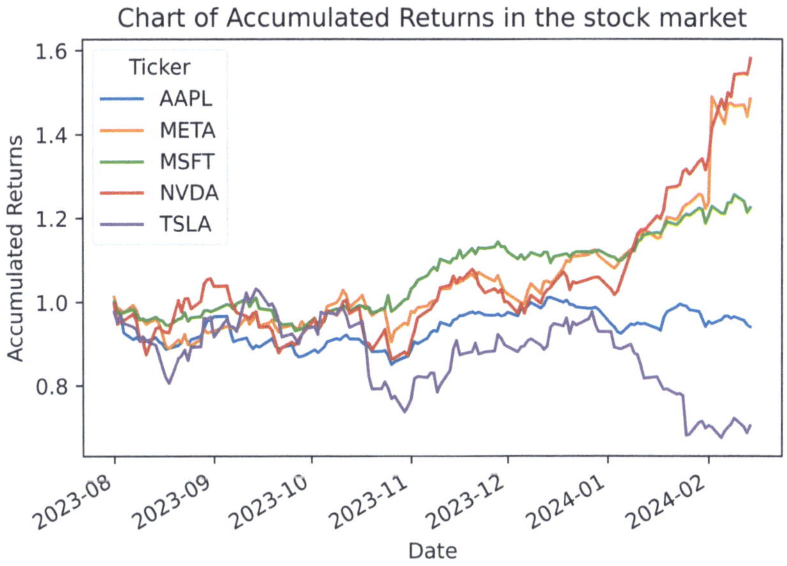

Fig. 3. Cumulative returns on selected stocks.

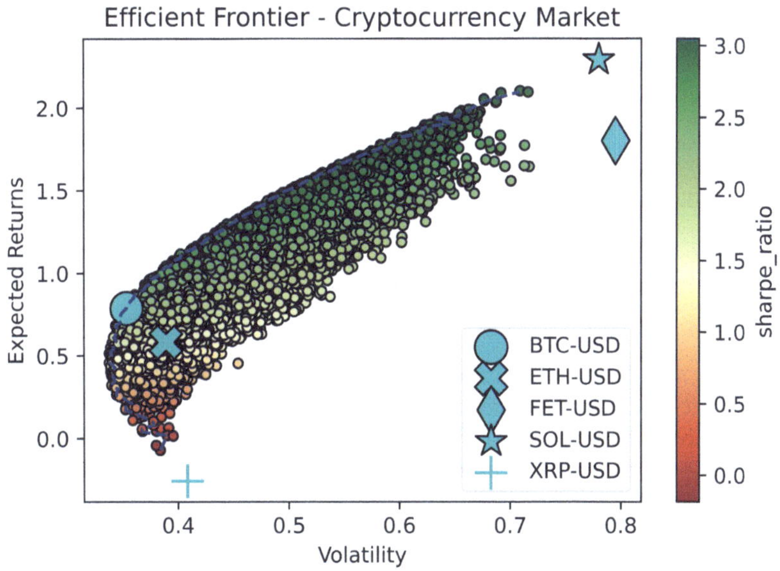

Fig. 4. Efficient frontier of minimum variance

the frontier, as these allow for a combination where the risk is minimized without sacrificing the portfolio's profitability.

Finally, if the user does not choose the Markowitz model, they will be able to compare the performance of purchasing individual assets against implementing strategies such as VixFix, moving averages, or momentum. For example, Fig. 5 shows how, using the VixFix model and the S&P500, an impressive accumulated return can be achieved. This is because, without this model, accumulated returns could be underestimated. In this sense, the results obtained show a thorough analysis of the cumulative return of the mentioned markets and their volatilities. These are displayed graphically on the screen and are easy to understand.

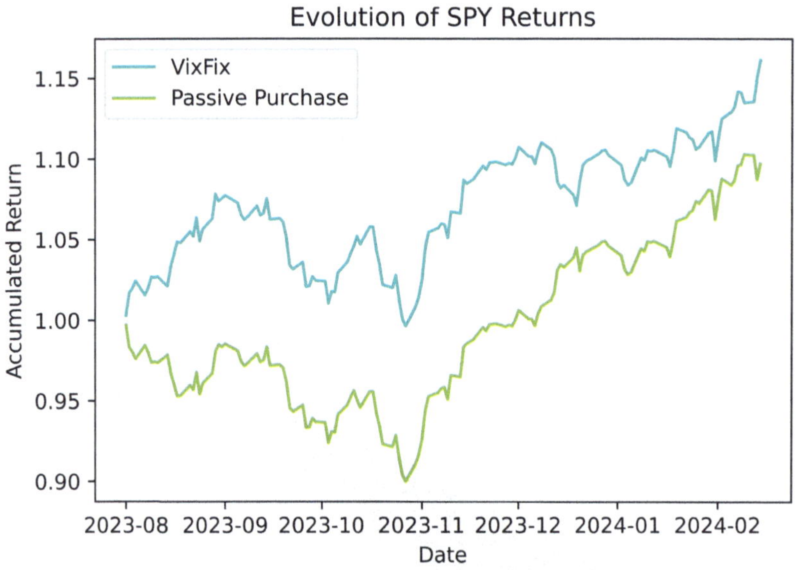

Fig. 5. Accumulated returns under alternative models

Although this ETF experienced declines in November 2023, using VixFix as an investment strategy would have allowed the portfolio to stabilize without incurring losses and take much better advantage of this asset's correction compared to passive buying. The user has the opportunity to test these strategies and evaluate which assets perform better under them.

This is how a system was implemented that, through user interaction, allowed various technical analysis strategies to be recreated to help underpin investment strategies. The user has the opportunity to choose the system and the asset from the range of possibilities offered. The results are immediate and the outcomes are descriptive analytical models presented in charts, making them easy to understand and summarizing a lot of information in intelligible graphs.

In the literature, there are studies that aim to address the prediction of financial asset prices using other learning methodologies. For example,

Pijus et al. (2022) implement predictive machine learning models in Python and demonstrate their feasibility [20]. Another example is Singh and Seth (2023), who use technical indicators and preprocess financial news to create market signals that predict market trends with high accuracy [26]. Unlike these articles, the present work includes algorithms that automate the reconstruction of various models and provides an interface that allows users to select the assets they wish to analyze.

6 Conclusion

A Python program has been developed with the main function of analyzing assets using financial models. This allows the user to choose from a range of assets and models to analyze their prices according to individual preferences, supported by real-time data from Yahoo Finance. In this way, the user can evaluate the results and choose the option that best suits their expected profitability and risk aversion level.

The project successfully implemented four different models, thus providing users with the freedom to select the market and specific asset they wish to analyze. This approach ensures the provision of customizable strategies and a variety of models to accommodate different needs. In addition, the program includes the presentation of real-time data through graphs, thereby facilitating visual assistance that significantly improves understanding of information and identification of capital market trends.

By incorporating financial models into the interactive program, it is expected that access to investment opportunities and financial education will increase among the population, especially in Latin America. This is due to the program's user-friendly interface, which allows users to experiment with different models for their chosen assets across various markets through graphs that enhance understanding of the capital market and enable the construction of an optimal portfolio.

However, the project has some limitations. Due to the relatively small size of local markets globally and limited user accessibility, structural issues may arise that could undermine the platform's reliability. Moreover, while assets inherently exhibit some volatility, their prices are highly susceptible to social, political, and economic conditions, especially in Latin America. Lastly, if users possess advanced financial knowledge and wish to adjust date ranges (omitting intervals within them) to analyze asset volatility, they would need to manipulate the Python code.

Finally, this project represents a significant first step towards opening up financial markets, thereby overcoming inefficiencies that limit development in the region. The use of Python to analyze assets using theoretical frameworks and real-time data collected from Yahoo Finance offers users the opportunity to identify investment opportunities and risks. Ultimately, the platform created is a valuable resource for students and financial enthusiasts alike, providing a tool for making informed decisions in the capital market, which at scale can contribute to economic development.

7 Recommendations and Future Work

There are very few articles that test the performance of assets under any investment strategy. It would be interesting to test more of these strategies where not only a single asset is tested, but an entire investment portfolio. Strategies that combine the purchase of equity assets with fixed income or even derivatives could also be included.

One recommendation that might be seen as a limitation in this work is to allow the user to manipulate the time frame in which the assets are analyzed. It would be interesting to design a console that gives the user the free opportunity to combine more assets from different sectors and to decide if they want to test a strategy over a different period than the 200 days.

This type of work has great potential in price analysis through the training of learning models. Once the initial prototype of the project is developed, it would be desirable to publish it so that other users can access a broader range of assets and financial models that are useful for making investment decisions. This way, the proposal can be validated by incorporating the necessary feedback that emerges afterward.

It would also be interesting to link the design of investment strategies with a stockbroker to effectively execute buy orders that meet the technical models recreated in this paper or others. In this way, investment strategies would be automated and investment decisions would be made in the shortest possible time. Assets could be classified under the strategies in which they perform best, creating an opportunity for people to invest with strategies that demonstrate their efficiency over time.

References

1. Ablanedo, G., Azcárraga, L., Mejía, E.: Moving averages as investment alternatives in a stock index. Horizontes de la Contaduría en las Ciencias Sociales (10), 38–57 (2019). https://www.uv.mx/iic/files/2020/01/Num10-Art03.pdf
2. Ananchotikul, S., Eichengreen, B.: Plumbing for Latin American capital markets. Bank Int. Settl. (36), 110–139 (2008). https://www.bis.org/publ/bppdf/bispap36.pdf
3. Banco Central de Reserva del Perú. Methodological Guide to the Weekly Note: Capital Markets. https://www.bcrp.gob.pe/docs/Publicaciones/Guia-Metodologica/Guia-Metodologica-04.pdf
4. Cooper, M.: Effectiveness of Technical Analysis versus Passive Position in the Retail Index in Chile [Undergraduate thesis, University of Chile]. Academic Repository of the University of Chile (2013). https://repositorio.uchile.cl/handle/2250/115036
5. Etoro. Target Market Determination CopyTrade (2022). https://www.etoro.com/wp-content/uploads/2022/05/TMD_CopyTrade.pdf
6. European Central Bank. Financial Stability Review (2007). https://www.ecb.europa.eu/pub/pdf/fsr/financialstabilityreview200712en.pdf

7. Garcia, N., Grifoni, A., López, J., Mejía, D.: Financial education in Latin America and the Caribbean. Current Situation and Perspectives. Corporación Andina de Fomento (2013). https://www.oecd.org/daf/fin/financial-education/OECD_CAF_Financial_Education_Latin_AmericaES.pdf
8. Giner, J., Morini, S.: The VIX Index for Volatility Prediction: An International Study. Repository of the Department of Financial Economics and Accounting, University of La Laguna (2004). https://accedacris.ulpgc.es/bitstream/10553/916/1/688.pdf
9. Gómez-Cambronero, E.: The Momentum Factor: Study of a Unifactorial and Multifactorial Strategy in Low Capitalization Assets. Repository of the Faculty of Economic and Business Sciences, Comillas Pontifical University (2020). https://repositorio.comillas.edu/jspui/bitstream/11531/40717/1/EL%20FACTOR%20MOMENTUM%20Estudio%20sobre%20una%20estrategia%20unifactorialbaja%20capitalizacion-Gomez-Cambronero%20Moreno%20de%20Tejada%2C%20Elisa.pdf
10. Grajales, D.: Portfolio Management. A Critical Look Beyond Markowitz. AD-minister, 15 (2009). http://www.redalyc.org/articulo.oa?id=322327246008
11. Grupo Sura. Political Stability and Integration, Keys to Developing Capital Markets in Latin America (2023)
12. Hilpisch, Y.: Python for Finance: Mastering Data-Driven Finance. O'Reilly Media (2014)
13. Lintner, J.: The valuation of risk assets and the selection of risky investments in stock portfolios and capital budgets. Rev. Econ. Stat. **47**(1) (1965). https://doi.org/10.2307/1924119
14. Ma Weiming, J.: Mastering Python for Finance. Packt Publishing (2015). https://www.palmislandtraders.com/econ136/mpff.pdf
15. Markowitz, H.: Portfolio selection. J. Finan. **7**(1), 77–91 (1952). https://doi.org/10.2307/2975974
16. Mendizábal, A., Miera, L., Zubia, M.: The Markowitz model in portfolio management. Cuadernos de Gestión **2**(1) (2002). https://core.ac.uk/download/pdf/6565186.pdf
17. MSCI. Factor Focus: Momentum. https://www.msci.com/documents/1296102/1339060/Factor+Factsheets+Momentum.pdf
18. Mossin, J.: Equilibrium in a capital asset market. Econometrica **34**(4) (1966). https://doi.org/10.2307/1910098
19. Murphy, J.J.: Technical Analysis of the Financial Markets (1999)
20. Pyke, P., et al.: Stock Price Prediction using Machine Learning with Python (2022). https://ijirt.org/master/publishedpaper/IJIRT155434_PAPER.pdf
21. TradeStation. Learning EasyLanguage Essentials (2017). https://uploads.tradestation.com/uploads/EasyLanguage-Essentials.pdf
22. Tran, M.: Yahoofinance Documentation (2020). https://python-yahoofinance.readthedocs.io/_/downloads/en/stable/pdf/
23. Raccanello, K., Herrera, E.: Education and financial inclusion. Revista Latinoamericana de Estudios Educativos **44**(2), 119–141 (2014). http://www.redalyc.org/articulo.oa?id=27031268005
24. Rodríguez, Á.: The Capital Market as an Investment Alternative. Visión Gerencial (2007). http://www.redalyc.org/articulo.oa?id=465545877012
25. Sanchez, A.: Asymmetric Information in the Capital Market: Colombia [Degree thesis, Pontifical Xavierian University]. Academic Repository of the Pontifical Xavierian University (2015). https://doi.org/10.11144/Javeriana.10554.18492

26. Singh, C., Seth, J.: Stock Market Prediction Using Python (2023). https://ssrn.com/abstract=4444871
27. Sharpe, W.F.: Capital asset prices: a theory of market equilibrium under conditions of risk. J. Financ. **19**(3), 425–442 (1964)
28. Williams, L.: Long-Term Secrets to Short-Term Trading. Wiley Trading Advantage (1999)

Software Engineering

Dynamic Test Case Prioritization and Selection for Continuous Integration Using Reinforcement Learning

Saad Waseem, Aicha Moussaid, Ricardo Chavez Tapia(✉), S. M. Zahid Hasan, Iván Porres Paltor, and Sebastien Lafond

Faculty of Science and Engineering, Åbo Akademi, Tuomiokirkontori 3, 20500 Turku, Finland
{saad.waseem,aicha.moussaid,ricardo.chaveztapia, zahid.hasan,ivan.porres,sebastien.lafond}@abo.fi

Abstract. Regression testing is an integral part of the continuous integration (CI) environment that gets executed as part of the CI cycle to ensure software system stability. It involves test case selection and prioritization steps that are required to meet the resource and time constraints within a CI cycle. This paper introduces dynamic test case prioritization and selection (DTPS), a novel test case selection and prioritization approach that performs effectively in a CI environment and minimizes regression time, hence providing early developer feedback to failed test cases. Reinforcement learning is used by DTPS to prioritize test cases based on their previous execution, i.e., historical runs, incorporating the latest acquired knowledge about failed test cases as the test cases are executed through a verdict-based reward function. Test volatility is also accounted for by DTPS in the CI context, where tests can be added/removed over CI cycles. Finally, DTPS is compared to the state-of-the-art reinforced test case selection (RETECS) method and two heuristic methods, over a recently published industrial dataset—Westermo (https://github.com/westermo/test-results-dataset)—to demonstrate its practical efficacy in a CI environment. The results show that DTPS outperforms RETECS, with a comparatively close schedule when compared to heuristic approaches after the initial exploration phase.

Keywords: Test case prioritization · continuous integration · regression testing · test case selection · reinforcement learning · machine learning

1 Introduction

Continuous integration (CI) is standard in the software industry, where developers frequently integrate code changes, triggering automated builds and tests to reduce integration issues and accelerate development [5,6]. While CI ensures early feedback, it challenges regression testing, which is crucial for ensuring that

F. Suni-Lopez et al. (Eds.): CIIS 2024, ISEM 48, pp. 149–163, 2025.
https://doi.org/10.1007/978-3-031-91481-2_10

changes don't negatively impact existing functionalities [2]. The dynamic nature of CI-with frequent updates and limited test budgets-renders traditional regression testing methods impractical for large codebases [4,5,7].

To address this, test case selection and prioritization (TSP) techniques have emerged, aiming to efficiently detect faults in new changes while providing early feedback. However, TSP methods must be fast and efficient to meet CI's dynamic nature and time constraints. Traditional TSP methods often rely on data sources like coverage analysis and test execution history [8,14], while recent approaches use machine learning (ML) to predict faults. Despite their potential, most ML techniques lack continuous adaptation and struggle in CI's dynamic environment [3,11]. Conversely, reinforcement learning (RL) has shown promise in adaptive scenarios like TSP, due to its ability to learn and optimize from interactions with the environment [1,3,15]. In this paper, we introduce a novel method that prioritizes tests one-by-one using RL, optimizing test case prioritization (TCP) by leveraging recent experience.

The main contributions of this work are as follows:

- We introduce dynamic test case prioritization and selection (DTPS), an RL-based method that prioritizes test cases one-by-one as they are executed, shifting the research focus from static to dynamic approaches in CI regression testing.
- DTPS meets CI's time and resource constraints, outperforming random and reinforced test case selection (RETECS) models to efficiently prioritize test cases with just two months of data. It also achieves results comparable to heuristic methods.
- Our experiments use the same industrial datasets as RETECS, in addition to a recently published dataset that was converted to meet the needs of DTSP, also available for future research. This demonstrates our approach's applicability in modern industrial settings and its language-agnostic nature.

The paper is organized as follows: Sect. 2 presents the DTPS approach, covering RL and TSP concepts. Section 3 outlines the research questions and evaluation method. Results and analysis are in Sect. 4, with related work in Sect. 5. Section 6 concludes the paper.

2 The DTPS Method

This section introduces our approach to the TSP problem using RL, known as DTPS. It presents two main points: first, in Sect. 2.1, a description of how RL is applied to TSP, and second, in Sect. 2.2, a discussion of the reward function.

2.1 Test Case Prioritization with Reinforcement Learning

This section describes the main components of RL for test case prioritization and selection, focusing on a model-free approach with online learning for the DTPS problem. The agent prioritizes each test case based on its importance before generating a test schedule. A model-free approach, which lacks initial knowledge of the environment, is suitable for DTPS since software failures and their detection do not follow a strict model. Figure 1 illustrates the link between RL and TCP. A state consists of test case's approximated duration, the time it was last executed, and its previous test execution results, denoted as t_i in DTPS Algorithm 1.

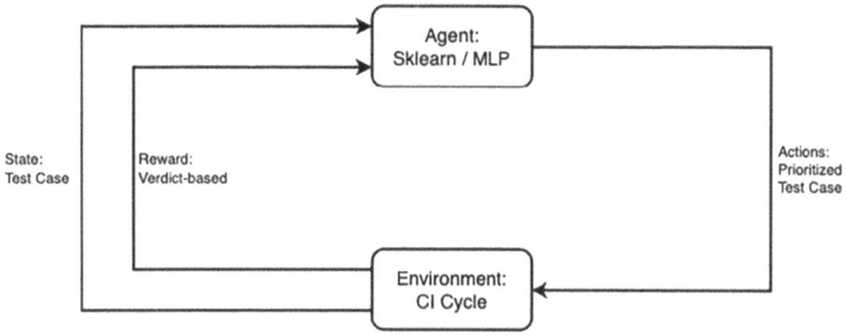

Fig. 1. Interaction of RL Agent and CI Environment (adapted from Kim and Porter [8] Fig. 1)

During the training phase (signaled by the variable *trainMode* having the value True in Algorithm 1), the agent assigns test case priorities using a policy function based on an artificial neural network (ANN) (*Multilayer Perceptron*), which maps the state (t_i) to an action, i.e., how important each test case is for the current CI cycle. Initially, this function (*AssignPriority*) approximates the optimal policy; it then gradually adapts towards an optimal policy with the accumulation of experience, driven by the reward function (see Sect. 2.2). Each experience consists of a test case, assigned priority, and reward, stored in a replay memory with a limited capacity; if the replay memory hits its limit, the oldest experiences are replaced first. In the training phase, random batches from this memory are used to train the ANN through backpropagation with stochastic gradient descent [10].

The ANN's architecture corresponds to one input layer, one hidden layer with 12 nodes—increasing the number of hidden nodes beyond 12 did not significantly improve performance [15]—and a single output layer. Table 1 gives an overview of the selected parameters based on values from the literature and experimental exploration.

During early epochs, the agent takes random actions, derived from a Gaussian distribution, to explore the environment and understand the effects of its actions by observing the resulting rewards. The exploration rate, which controls the frequency of random actions, starts high to encourage exploration and then decreases over time as the agent begins relying more on its learned policy. However, exploration continues throughout learning since the dynamic environment and non-stationary rewards require the policy to adapt continuously through online learning.

Table 1. MLP Parameters

Parameter	Value
CI cycle's time limit	$80\% \times (TotalTimeofCI)$
History length	4
Hidden nodes	12
Replay memory	10000 [15]
Replay batch size	1000 [15]

The process of test case selection follows a rather simple approach. First, the prioritized test cases are arranged in descending order based on their significance. Second, for each prioritized test case, its approximated execution duration is extracted from the previous CI cycle and aggregated with the ongoing accumulation of execution time (referred to as *executionTime* in Algorithm 1). When this accumulated execution time aligns with the designated time budget, the schedule is finalized. Subsequent evaluation of the produced schedule is performed, employing the normalized average percentage of faults detected(NAPFD) metric as delineated in the evaluation framework provided in Evaluation Setup (see Sect. 3).

2.2 Reward Function

In the context of DTPS, the effectiveness of a test schedule is measured by its ability to detect failures at an early stage, leading to reduced regression testing time. Therefore, the test schedule must comprise test cases that have a higher probability of identifying failures early on. To ensure that the agent takes appropriate actions, it is crucial to incorporate these objectives and domain expertise into the reward function [9].

In DTPS, the reward function evaluates test case verdicts and provides feedback to the agent with a higher positive or negative magnitude. Higher positive magnitudes steer the agent to prioritize revealing test cases over non-revealing test cases, hence optimizing failure detection and reducing regression time. For the test case schedule evaluation, a common metric, NAPFD, (presented in Sect. 3.2) is used. However, these metrics require knowledge about the total number of faults in the system under test or full information on test case verdicts, even for non-executed test cases.

$$\text{reward}_i^{tcfail} = \begin{cases} t.\text{verdict}_i \times 99999 & \text{if } t \in T_s \text{ and } t.\text{verdict}_i = 1 \\ -99999 & \text{otherwise} \end{cases} \quad (1)$$

In a CI environment, the verdict of a test case can only be obtained after the test has been executed and information about missed failures is not available. An RL agent can only learn to reinforce actions that have shown positive results. It can not learn about test cases that should have been included. Therefore, in

Algorithm 1. DTPS Algorithm

```
Require: Test suite T_s = {t_1, ..., t_n}
  procedure DTPS(T_s)
    for t_i ∈ T_s do
      // Perform operations on t_i
      // Assign priorities, calculate rewards, update model, etc.
      priority ← AssignPriority(t_i)
      if trainMode = True then
        verdict ← ExecuteTestCase(t_i)
        reward ← CalculateReward(verdict)
        // For experience replay
        StoreRewardInMemory(t_i, reward)
        if iterationCount mod 5 = 0 then
          UpdateModel(s_n)
        end if
      end if
      Add (priority, t_i) to prioritized list
    end for
    availableTimeBudget ← 0.8 × totalDuration(for T_s)
    Sort prioritized list by priority in descending order
    selectedTestCases ← empty list
    currentTime ← 0
    for priority, t_i in prioritized list do
      executionTime ← EstimateExecutionTime(t_i)
      if currentTime + executionTime ≤ availableTimeBudget then
        Add t_i to selectedTestCases
        currentTime ← currentTime + executionTime
      else
        break
      end if
    end for
    napfdValue ← CalculateNAPFD(selectedTestCases)
    return selectedTestCases, napfdValue
  end procedure
```

DTPS, rewards are either positive or negative based on the verdict of executed test cases, as defined by Eq. (1).

In order to teach the agent about both behaviors, we implemented an exhaustive approach and executed all prioritized test cases to enrich the agent's experience with both scenarios. This only happens during training: at inference time, the agent prioritizes test cases and then follows a selection strategy before test cases are executed. The agent gets a positive reward for detecting failure, and a negative reward, or penalty, for executing a passed test case. The order of test cases is not explicitly included in the reward. It is implicitly included by encouraging the agent to focus on failing test cases and prioritizing them higher. For the proposed scheduling method, this may lead to an earlier execution of failed test cases.

3 Evaluation Setup

In this section, we present the evaluation setup adopted to test our approach, along with an explanation for the chosen evaluation metric. Initially, we state three research questions to evaluate the proposed approach followed by a brief introduction of the metric used in Sect. 3.1. The experimental setup is then explained in Sect. 3.2; lastly, a detailed layout of the datasets used is presented in Sect. 3.3.

- RQ1: Can DTPS perform at par with deterministic methods? We compute NAPFD and compare its mean value for every 30 CI cycles over a dataset of 210 CI cycles.
- RQ2: Can DTPS perform comparably to RETECS using verdict-based reward? We compare DTPS and RETECS using various datasets having varying failure ratios.
- RQ3: Is DTPS effective in prioritizing and selecting test cases? We analyze the overall efficiency of DTPS and its deployability in industrial settings.

3.1 Evaluation Metric

To evaluate our approach, we used normalized average percentage of faults detected average percentage of faults detected (NAPFD) (Eq. 2) introduced by Qu et al. [13] as an extension of APFD (average percentage of faults detected) to account for the ratio between detected and total failures in a test suite. NAPFD is more suitable for scenarios where not all test cases are executed, and some failures may remain undetected. If all faults are detected ($p = 1$), NAPFD equals the original APFD.

$$\text{NAPFD} = p - \frac{sum(D_{\text{ranks}})}{tf \times T_s} + \frac{p}{2 \times T_s} \tag{2}$$

$$p = \frac{df}{tf} \tag{3}$$

where D_{ranks} is the list of test case ranks within a CI cycle. T_s is the total number of test cases within a test suite. *df* is detected failures and *tf* is total failures (sum of verdicts).

3.2 Experimental Setup

We implemented an RL agent based on an ANN to process continuous states, i.e., test cases, and produce a single action, i.e., Priority. The agent used a reward function based on the test case verdict to guide the policy toward a better approximation of failure detection as shown in Fig. 2. Test cases are scheduled on a single test agent in descending order of priority until the time limit is reached.

For comparative analysis, we evaluate the efficiency of the DTPS method by comparing it to four different TCP methods: a baseline random model, two deterministic methods (sorting and weighting), and the state-of-the-art RETECS approach. The sorting method prioritizes test cases based on recent failures, while weighting assigns equal importance to all test case attributes. The RETECS approach, on the other hand, generates a test schedule and collects rewards before providing feedback to the agent. All methods used the same information encountered in industrial environments, and performance was assessed using the NAPFD metric across 210 CI cycles with at least a 10% failure rate, as our primary interest lies in evaluating how well these methods perform under such circumstances. Each experiment was repeated 30 times to account for variability in the results, allowing us to produce averaged results that reflect consistent patterns rather than isolated occurrences.

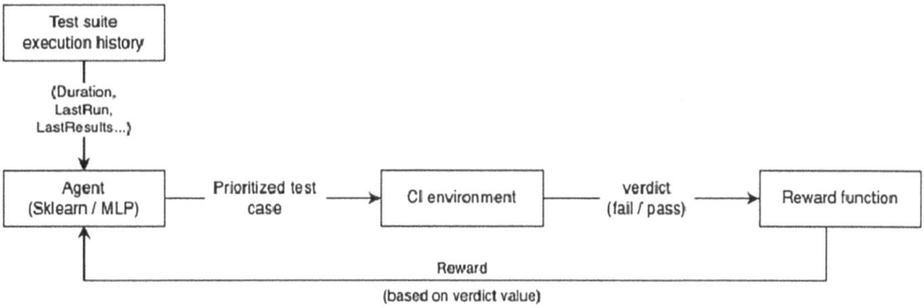

Fig. 2. Reference Architecture - Dynamic Test Case Prioritization and Selection using Reinforcement Learning (Training Phase)

3.3 Industrial Datasets

We conducted experiments on three datasets, including two used by RETECS [15]. Industrial datasets from ABB Robotics Norway[1], *Paint Control*, and *IOF/ROL* for advanced industrial robot testing, contain data for over 300 CI cycles across daily intervals, with historical test execution and verdict data. The Westermo dataset, crucial for DTPS and TSP research, includes over one million verdicts from embedded system testing across 500 consecutive days [16]. The dataset structures are summarized in Table 2.

It is worth noting that the Westermo dataset required customization to align with our approach which involved combining attributes and understanding Westermo's testing process. The resulting dataset can be freely retrieved from Zenodo[2]. This conversion allowed us to perform a detailed comparative analysis of

[1] Website: http://new.abb.com/products/robotics.
[2] https://zenodo.org/record/7941024.

Table 2. Industrial Datasets Overview

Datasets	Test Cases	CI cycles	Verdicts	Failed
ABB Paint Control	89	352	25594	19.36%
ABB IOF/ROL	1,941	320	30,319	28.43%
Westermo	1,855	15,197	1,036,818	5.03%

deterministic and RL-based TSP methods (RETECS & DTPS) under low failure rates. Section 4 presents detailed analysis and challenges posed by limited availability of datasets for RL-based TSP research.

4 Results and Analysis

In this section, we present and analyze the obtained results. The first two questions compare the performance of DTPS using a well-known metric, NAPFD, over three datasets, shown in Table 2, with varying failure rates. The third question summarizes the results of our analysis and findings.

4.1 RQ1: Can DTPS Perform at par with Deterministic Methods?

RQ1's focus is on comparing DTPS with two heuristic methods: sorting and weighting. Heuristic sorting orders the test cases based on their latest verdicts, with failed test cases being sorted first; if two test cases have the same verdicts in their latest runs, the most recently executed one is sorted first. On the other hand, heuristic weighting sorts the test cases according to the weighted average of the elements of the feature vector; the default approach is to give all features equal weight, which would produce a regular average.

In order to draw a thorough comparison, all three methods are evaluated over the three previously mentioned datasets using NAPFD values. Figure 3 shows the results of this comparison over the Westermo dataset. During the initial phase, the deterministic methods performed better, since this corresponds to the exploration phase for DTPS where the method is trying to adapt to its environment. For the Westermo dataset, after 60 CI cycles, DTPS showed a relative increase in average NAPFD values and prioritized the test cases with similar performance to comparison methods.

IOF/ROL exhibited similar results, albeit with diminished performance, while Paint Control demonstrated an initial comparable performance in terms of NAPFD that gradually declined in comparison to heuristic methods in subsequent CI cycles. Figure 4 provides a summary of the comparison for the aforementioned datasets.

For IOF/ROL, where DTPS showed lower performance, heuristic methods were also not able to prioritize failing test cases higher than the passed ones, as indicated by the narrow performance gaps in the left plot of Fig. 4.

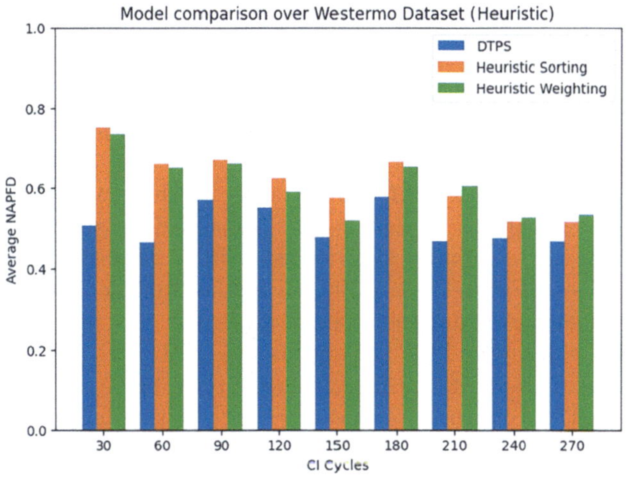

Fig. 3. Comparison of DTPS results against heuristic methods over Westermo dataset

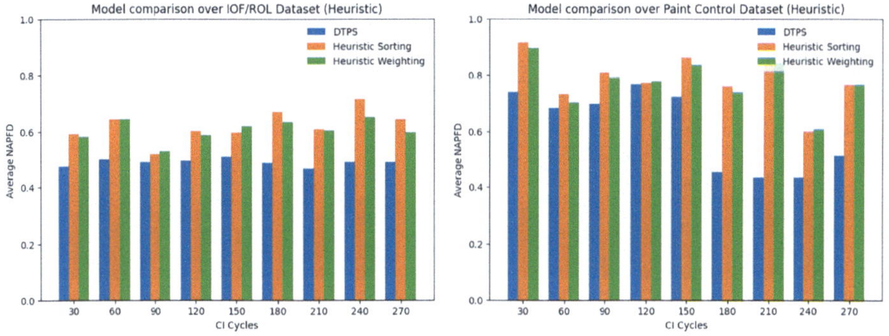

Fig. 4. Comparison of DTPS and heuristic methods over IOF/ROL and Paint Control datasets

For Paint Control, DTPS achieves early comparable performance to heuristic methods due to the presence of higher failures in initial CI cycles, which results in quick adaptability of the agent and better results as shown in the right plot of Fig. 4. As the agent continues with later CI cycles, the performance drops due to fewer failures within the chosen batch of 30 CI cycles. Also, it appears that the later CI cycles consist of test cases that are not as correlated (history length was insignificant) to the previous test execution results as the former CI cycles.

In summary, the results for RQ1 show that, despite the initial low performance of DTPS compared to the deterministic methods, it exhibits promising adaptability and competitiveness as the CI cycles progress, especially with higher test failure rates. This highlights the potential DTPS holds as an effective alternative to deterministic methods in the CI context.

4.2 RQ2: Can DTPS Perform Comparably to RETECS Using Verdict-Based Reward?

Before presenting comparison results of RETECS with the *tcfail* reward function[15] and DTPS, it is crucial to understand the fundamental difference between the two methods.

The former method based its reward strategy on the generated test schedule, which means the agent has to wait for the test schedule to get its reward and then incorporate the generated knowledge in order to perform better for the following CI cycles. However, the proposed method reduced the feedback cycle to one test case and provides the most recent experience to the agent leading to an increased efficacy of DTPS. Furthermore, in DTPS, the agent's reward strategy integrates test case verdicts, thereby providing positive feedback for executing a failed test case and a negative reward for running a passed test case. This helped the agent learn a policy that strives to run failed test cases to reduce regression time in a CI context.

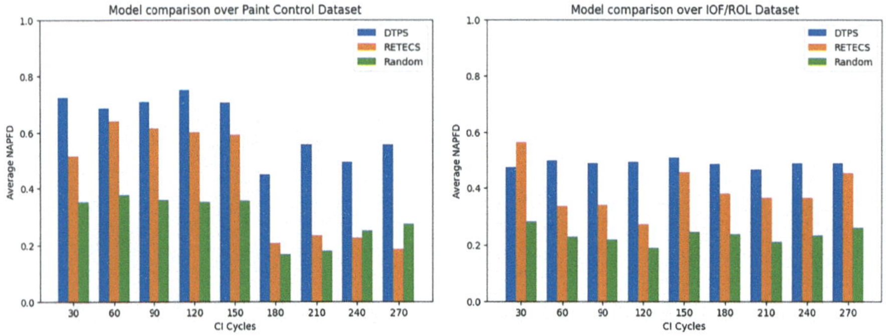

Fig. 5. Comparison of DTPS with RETECS and Random Models over Paint Control

Paint Control has a higher percentage of failures in initial CI cycles as indicated in Fig. 5 by high NAPFD values for both agents, RETECS and DTPS. As the agents continue prioritizing test cases for following CI cycles that have a reduced number of failed test cases, RETECS was not able to perform comparably to DTPS as indicated by the decreasing mean values for NAPFD, whereas the proposed method performed consistently better than RETECS and random methods in those scenarios.

However, IOF/ROL has more failures than Paint Control with fewer historical runs, particularly for initial CI cycles. During the exploration phase, DTPS showed reduced performance in comparison to RETECS but after 30 CI cycles it adapts to the environment and learns to rank failed test cases higher as indicated by improved NAPFD values approaching 0.5. RETECS, on the other hand, had difficulty identifying a connection between failed test cases when there were insufficient historical runs available in the data set. This suggests that RETECS is more sensitive to the absence of past runs for failed test cases than DTPS.

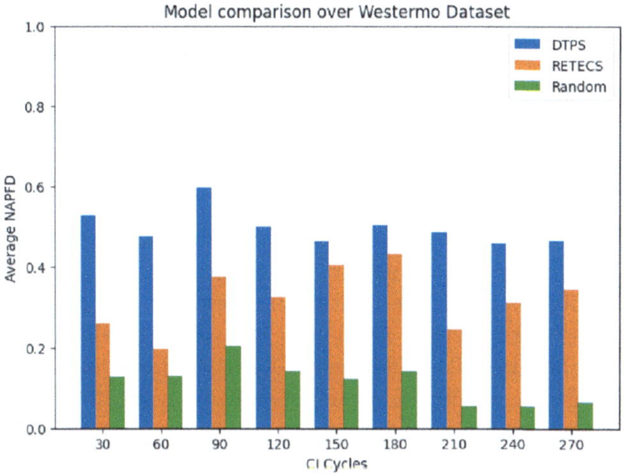

Fig. 6. Comparison of DTPS and RETECS over Westermo data set

In order to evaluate DTPS and RETECS performance in scenarios of reduced failure rate and frequent CI cycles with a significantly large number of test case verdicts, we executed similar experiments over the Westermo data set, which contains test execution history for 11 different software systems. Figure 6 presents the results of these experiments for System 11 since it has a relatively low failure rate with reasonable correlation among test cases. We found that DTPS performed consistently better than RETECS and random methods despite the reduced failure rate with frequent executions. Hence, we conclude that the DTPS reward strategy and experience replay is crucial to its robustness and demonstrated performance in such cases, which makes it more suitable for a CI environment where the test suite and test executions keep changing.

4.3 RQ3: Is DTPS Effective in Prioritizing and Selecting Test Cases?

Dynamic test case prioritization and selection incorporates a verdict-based reward strategy and provides immediate feedback to the agent enriching its experiential memory with a comprehensive overview of both failed and passed test cases. These capabilities ensure the robustness and efficacy of DTPS in a CI environment where test suite changes are frequent.

In this third and final case, we assessed DTPS's resilience to reduced failure rates with reasonable availability of historical runs, as exemplified by the Westermo dataset. Dynamic test case prioritization and selection consistently outperformed RETECS, demonstrating its ability to prioritize failed test cases in a dynamic environment, ultimately reducing regression testing time. However, in scenarios where both the failure rate and the availability of historical runs for failed test cases drop significantly, DTPS's performance may also be impacted.

The agent might struggle to find correlations among failed test cases, but it will still outperform RETECS, as shown in Fig. 6. Therefore, we conclude that DTPS is less sensitive to reduced failure rates, provided there are sufficient historical runs available, and it can outperform RETECS in the CI context.

4.4 Threats to Validity

In this section we discuss the main threats to the validity of our results and how we mitigate them, using the taxonomy of Wholin et al. [18].

Internal validity concerns the relationship between our treatment and outcome. Potential threats include implementation errors, which we mitigate by using established libraries like scikit-learn. Another threat is the parameter settings of ML algorithms, which might not generalize across different environments. To address this, we kept parameters consistent across all experiments, though they could be tuned for specific real-world scenarios.

Conclusion validity pertains to drawing accurate conclusions. In our case, randomness in results could threaten validity, so we repeated experiments 30 times and averaged the outcomes to minimize this risk. We also used well-established metrics, such as NAPFD, to ensure comparability and reliability of our conclusions [15].

External validity relates to the generalizability of our findings. Our study is based on three industrial datasets, two of which are from prior research. Although these datasets might not cover all CI environments and failure distributions, they provide a strong foundation. However, to address these concerns further experiments with more diverse datasets are recommended as they become available.

5 Related Work

The use of ML in test case selection and prioritization has been widely explored. Pan et al. [11] conducted a systematic review of 29 papers, highlighting that while most focus on supervised learning, these methods are not ideal for the dynamic nature of CI due to their lack of incremental learning. In contrast, RL techniques allow for the continuous integration of new data without retraining.

Bertolino et al. [3] categorized ML approaches into supervised learning (learning-to-rank) and RL (ranking-to-learn), recommending RL as better suited for CI environments. Bagherzadeh et al. [1] evaluated various RL approaches with different ranking models, showing that specific combinations can achieve high accuracy, particularly with datasets including code-related features, though these features can be language-dependent and less generalizable. Spieker et al. [15] introduced RETECS, an RL-based approach for TCP, which serves as a baseline for our work. We extend this by incorporating dynamic scheduling that takes immediate feedback after each test case, rather than waiting for the entire test suite to complete.

Further enhancements to RETECS include Wen et al.'s [17] use of FP-Growth for pattern-based reprioritization, which analyzes historical test execution data to adjust the RETECS-generated schedule. Similarly, Pradhan et al.'s

[12] REMAP employs rule—mining and search—based algorithms for dynamic scheduling, using multi-objective search to build an initial schedule and the RIPPER algorithm to reprioritize as test verdicts are determined.

Our approach differs by integrating recent test execution feedback on a per-test-case basis, reducing the agent feedback cycle and improving the prioritization of failed test cases, especially in low-failure-rate scenarios. Additionally, we compare DTPS and RETECS using average NAPFD values on a newer dataset, with a reduced failure rate compared to Paint Control and IOF/ROL.

6 Conclusion

This work presents and evaluates a novel approach to test case selection and prioritization that relies only on lightweight features that can be obtained from a test suite's execution history. Dynamic test case prioritization and selection uses a reinforcement learning agent to prioritize test cases one by one, as they are being executed, incorporating the most recent knowledge being produced during the CI cycle and significantly reducing the regression testing time. This is done with relatively higher accuracy than RETECS, our baseline method.

We evaluated our approach on two existing datasets, that were also used by RETECS and one recently published industrial dataset. Our experimental results show that DTPS can outperform RETECS and performs comparably close to heuristic methods, requiring only 2 months of test execution history to become efficient at prioritizing test cases. The method is model-free, language-agnostic, and requires no source code or program access. It only requires test metadata-namely historical results-test duration, and last execution timestamps. However, we expect additional metadata to enhance the method's performance. In our evaluation, we compared different failure ratios with a verdict-based reward function and found that the presence of a higher number of failing test cases leads to higher performance.

As future work, we intend to investigate how features like code complexity and code coverage could affect DTPS's ability to detect failures. We should also develop a new evaluation metric for assessing DTPS performance in varying CI setups.

Acknowledgments. The work has been partially supported by the EMJMD master's program in Engineering of Data-intensive Intelligent Software Systems (EDISS - European Union's Education, Audiovisual and Culture Executive Agency grant number 619819).

Data Availability Statement. The data described in the article is openly accessible on Zenodo at https://doi.org/10.5281/zenodo.8401356.

Declarations

Conflict of Interest. The authors have no conflict of interest.

References

1. Bagherzadeh, M., Kahani, N., Briand, L.: Reinforcement learning for test case prioritization. IEEE Trans. Softw. Eng. **48**(8), 2836–2856 (2022). https://doi.org/10.1109/TSE.2021.3070549
2. Basu, A.: Software Quality Assurance, Testing and Metrics. PHI Learning (2015). https://books.google.com/books?id=aNTiCQAAQBAJ&pg=PA150
3. Bertolino, A., Guerriero, A., Miranda, B., Pietrantuono, R., Russo, S.: Learning-to-rank vs ranking-to-learn: strategies for regression testing in continuous integration. In: Proceedings of the ACM/IEEE 42nd International Conference on Software Engineering, ICSE 2020, pp. 1–12. Association for Computing Machinery, New York (2020). https://doi.org/10.1145/3377811.3380369
4. Do, H., Mirarab, S., Tahvildari, L., Rothermel, G.: An empirical study of the effect of time constraints on the cost-benefits of regression testing. In: Proceedings of the 16th ACM SIGSOFT International Symposium on Foundations of Software Engineering, SIGSOFT 2008/FSE-16, pp. 71–82. Association for Computing Machinery, New York (2008). https://doi.org/10.1145/1453101.1453113
5. Duvall, P.M., Matyas, S., Glover, A.: Continuous Integration: Improving Software Quality and Reducing Risk. Addison-Wesley Professional, Boston (2007)
6. Fowler, M., Foemmel, M.: Continuous integration (2006). https://martinfowler.com/articles/continuousIntegration.html
7. Khatibsyarbini, M., Isa, M.A., Jawawi, D.N., Tumeng, R.: Test case prioritization approaches in regression testing: a systematic literature review. Inf. Softw. Technol. **93**, 74-93 (2018). https://doi.org/10.1016/j.infsof.2017.08.014
8. Kim, J.M., Porter, A.: A history-based test prioritization technique for regression testing in resource constrained environments. In: Proceedings of the 24th International Conference on Software Engineering, ICSE 2002, pp. 119–129. Association for Computing Machinery, New York (2002). https://doi.org/10.1145/581339.581357
9. Mataric, M.J.: Reward functions for accelerated learning. In: W.W. Cohen, H. Hirsh (eds.) Machine Learning Proceedings 1994, pp. 181–189. Morgan Kaufmann, San Francisco (1994). https://doi.org/10.1016/B978-1-55860-335-6.50030-1. https://www.sciencedirect.com/science/article/pii/B9781558603356500301
10. Noor, T.B., Hemmati, H.: A similarity-based approach for test case prioritization using historical failure data. In: 2015 IEEE 26th International Symposium on Software Reliability Engineering (ISSRE), pp. 58–68 (2015). https://doi.org/10.1109/ISSRE.2015.7381799
11. Pan, R., Bagherzadeh, M., Ghaleb, T.A., Briand, L.: Test case selection and prioritization using machine learning: a systematic literature review. Empir. Softw. Eng. **27**, 29 (2022). https://doi.org/10.1007/s10664-021-10066-6
12. Pradhan, D., Wang, S., Ali, S., Yue, T., Liaaen, M.: Remap: using rule mining and multi-objective search for dynamic test case prioritization. In: 2018 IEEE 11th International Conference on Software Testing, Verification and Validation (ICST), pp. 46–57 (2018). https://doi.org/10.1109/ICST.2018.00015
13. Qu, X., Cohen, M.B., Woolf, K.M.: Combinatorial interaction regression testing: a study of test case generation and prioritization. In: IEEE International Conference on Software Maintenance, 2007 (ICSM), pp. 255–264 (2007). https://doi.org/10.1109/ICSM.2007.4362638
14. Rothermel, G., Untch, R.H., Chu, C., Harrold, M.J.: Prioritizing test cases for regression testing. IEEE Trans. Softw. Eng. **27**(10), 929–948 (2001). https://doi.org/10.1109/32.962562

15. Spieker, H., Gotlieb, A., Marijan, D., Mossige, M.: Reinforcement learning for automatic test case prioritization and selection in continuous integration. In: Proceedings of the 26th ACM SIGSOFT International Symposium on Software Testing and Analysis, ISSTA 2017, pp. 12–22. Association for Computing Machinery, New York (2017). https://doi.org/10.1145/3092703.3092709
16. Strandberg, P.E.: The westermo test results data set. arXiv preprint arXiv:2211.13622 (2022). https://doi.org/10.48550/arXiv.2211.13622
17. Wen, W., Yuan, Z., Yuan, Y.: Improving RETECS method using FP-growth in continuous integration. In: 2018 5th IEEE International Conference on Cloud Computing and Intelligence Systems (CCIS), pp. 636–639 (2018). https://doi.org/10.1109/CCIS.2018.8691385
18. Wholin, C., Runeson, P., Host, M., Ohlsson, M.C., Regnell, B., Wesslén, A.: Experimentation in software engineering: an introduction. Massachusetts: Kluwer Acad Publ (2000). https://doi.org/10.1007/978-3-642-29044-2

Software Test Suite to Stress Non-Uniform Memory Access on Cluster Environments Using Cloud Technologies

J. A. Torres Ramos(✉)

Instituto Tecnológico y de Estudios Superiores de Monterrey (ITESM), Monterrey, Mexico
javier_alejandro_torres_r@hotmail.com

Abstract. Recent advancements in cloud technology have significantly enhanced the deployment and development of innovative technological solutions. Central to this progress are virtual machines optimized for high-performance tasks, utilizing non-uniform memory access (NUMA) technology to ensure consistent and reliable performance through low-latency, localized memory access for each central processing unit. This paper presents a comprehensive test suite specifically designed to evaluate NUMA's reliability across various stress scenarios. By leveraging cloud-based tools and implementing test case reduction strategies, the suite not only improves execution time efficiency, but also provides a detailed assessment of NUMA's operational resilience. Our findings demonstrate the effectiveness of the test suite in identifying and managing multiple uncorrectable memory errors, thereby validating its utility in optimizing testing processes. Notably, the application of this suite reduced the execution time of test cases from 24 to 14 h, marking a significant advancement in testing efficiency. This study underscores the potential of integrated NUMA technology in cloud environments to meet the demands of high-performance applications, highlighting substantial improvements in both efficiency and resource management.

Keywords: Non-Uniform Memory Access (NUMA) · Cloud Service Providers (CSPs) · Virtual Machines (VMs) · Docker · Kubernetes · Pairwise Testing

1 Introduction

In the last decade, the architecture of systems within cloud service providers (CSPs) has undergone significant evolution, especially around multiprocessor servers. One of the most important features of these developments has been non-uniform memory access (NUMA). This technology is especially relevant as it provides consistent and reliable performance of virtual machines (VMs) where it is crucial to meet user demands.

The motivation for the generation of this research lies in sharing and creating a tool to test some kernel features through a cloud native environment. First, problems arise from the complexity of validating these NUMA features on x86 architecture servers through virtual machines in a large number of systems.

F. Suni-Lopez et al. (Eds.): CIIS 2024, ISEM 48, pp. 164–177, 2025.
https://doi.org/10.1007/978-3-031-91481-2_11

Given this problem, a test suite designed to address the critical functionalities associated with NUMA in cloud environments has been developed. This suite aims to exercise local and remote memory access in systems across cluster environments, in addition to applying stress to processors and performing cache coherency tests. This is all carried out directly on the system, and the layer of virtualization is added through Kernel-based Virtual Machine (KVM).

To increase the versatility of the tool, several cloud technologies have been incorporated, including containers, VMs, and orchestrators. These technologies provide portability, environment emulation, and efficient test execution. Among the tools used are Docker, Kubernetes, and quick emulator (QEMU). The integration of these tools represents an advantage in addressing the challenges associated with test execution. The study proposes to explore in detail the design, implementation, execution strategy, and results obtained with the test suite, evaluating the effectiveness to ensure maximum quality, functionality, and reliability of NUMA features. Through this analysis, it seeks to contribute a valuable tool for CSPs in their effort to improve the end-user experience in the cloud. To achieve the objective of this study, the research and the project were conducted through a mixed methodology where the results of two experiments that were carried out are described in a quantitative way, and the qualitative part is seen only tangentially through multiple interviews to know the effectiveness, reliability, and usability of the tool.

In turn, the following research questions were posed:

- RQ1. To what extent is the test suite effective in detecting and diagnosing NUMA-related faults in multiprocessor systems?
- RQ2. How do cloud technologies, such as containers, VMs, and orchestrators influence runtime efficiency?

Finally, as a result, it can be known that the tool truly validates NUMA features such as local and remote access through bare metal systems, and virtualization with cloud native tools where two uncorrectable memory errors were found in two of the seven systems.

2 Background

2.1 Non-Uniform Memory Access (NUMA)

"Non-uniform memory access, better known as NUMA, is commonly found in multi-socket architectures. It also belongs to the category of shared memory architectures, where system resources such as physical processors, available dynamic random-access memory (DRAM), and the last level cache are divided into so-called NUMA nodes. These nodes are interconnected through a data bus" [1–3]. NUMA has local and remote memory accesses. "Local access occurs when the threads of a NUMA node access the memory resources of that same NUMA node to do so in a faster way and achieve low latency in the system" [4, 5].

On the contrary, "remote accesses are manifested in situations where threads belonging to a NUMA node access memory resources belonging to another NUMA node; in these situations, the system latency increases considerably" [6]. "This type of access is called hop 1, when it is interconnected with the other NUMA node that contains the memory accesses. On the other hand, hop 2 access is when the nodes are not directly interconnected" [7].

Within this topic, there are certain restrictions on the configuration of the systems so that the NUMA options can be used. Firstly, "there is the case of the memory configuration, here the system must be properly populated with valid configurations such as the 1DPC (DIMMS per channel) or 2DPC configuration" [8]. Subsequently, to enable this option, "it is necessary to adjust the memory scaling settings in the basic input/output system" [9,10]. In this section, several memory interleaving alternatives are presented, including the default settings. Among these options, it is common to find the possibility of configuring two NUMA nodes per socket.

Among the many existing pieces of research about NUMA, the authors perform benchmarking of the accesses by means of multiple tools or algorithms based on specific memory characteristics.

2.2 Cloud Technologies

Cloud technologies within this work play a very important role in performing the implementation and execution in an efficient way. In addition, these technologies can generate portability of dependencies, packages, and elements necessary for the test suite and try to emulate a cloud-like environment.

Firstly, within the cloud technologies used, the use of QEMU KVM is essential to carry out virtualization as it is done in the cloud, since with this tool, "we can customize the VMs in a similar way that CSPs do" [11].

Secondly, Docker containers within this work have were focused on generating portability and scalability. While it is a virtualization tool, it is used as a tool to be able to have the necessary elements needed by the test suite. "The generated Docker images could be executed, stopped, or destroyed, and serve to store and encapsulate applications or services along with the necessary dependencies to carry out the execution" [12,13]. Within the researched literature, some authors also "make use of containers as a tool to generate the portability of the applications where they inject stress to their systems" [14]. In turn, one of the reasons why they are quite useful is that they can share multiple layers such as the kernel, or the hardware layer.

Finally, "Kubernetes, the fundamental tool to carry out the execution, is an orchestration application. Here we orchestrate containerized solutions in cloud native microservice applications" [15]. "Some authors have created clusters based on Kubernetes where they make a comparison of lightweight versions of this application" [16]. In this case, they are relevant since they follow the fundamentals of cluster construction, following the architecture as in this study [17].

2.3 Pairwise Testing

Pairwise testing implementation is based on the need to reduce the number of test cases in the most efficient way, to prolong the execution time of each stressor and increase the chances of finding bugs.

There are studies that compare the number of test cases between ad-hoc and pairwise techniques to find the same number of issues. An example can be that "while ad-hoc techniques took 1090 tests to find 10 software issues, pairwise testing did it in only 19 executed test cases. In conclusion, running with pairwise versus conventional techniques reduced the effort by up to five times in terms of execution time, effort evaluating results, and person-hours" [18].

2.4 Stress and Benchmarking

The stress and benchmarking of the systems is the driving force of the project, in this field multiple investigations have been carried out.

Firstly, various tools, including stress-ng, Unixbench, Sysbench, stress-apptest, mprime, SPEC and iperf3, have been employed for use on systems with processors of different architectures, as well as on cards and virtual machines. The purpose of this implementation is twofold: on the one hand, to provoke hardware errors in a controlled manner and, on the other hand, to carry out detailed comparisons through the benchmarking process [19–21].

3 Methodology

The test suite aims to push NUMAs to the limit, for this purpose, this research uses a mixed methodology where both quantitative and qualitative data are obtained to validate the usefulness of the tool, where it should be noted that the qualitative part is only touched tangentially, and the focus of the study is observed quantitatively. On the other hand, the difference of this test suite against compared with other alternatives is found in the generation of a tool to validate NUMAs through multiple tools, either to generate stress, or to try to emulate a CSP as much as possible through NUMA nodes attached to the virtual machines where the systems will be exercised.

First, to obtain quantitative data, a total of two experiments were carried out where a system was tested to detect errors, detect bugs, and verify the effectiveness of the tool. In addition, the execution times were verified through Kubernetes and the manual execution of the containers to account for efficiency. On the qualitative side, interviews were conducted with a sample of 10 system validation engineers between 25 and 35 years old with more than 3 years of experience in the field to conclude if the test suite shows any kind of usefulness in the work environment. Figure 1 illustrates in detail the process followed for the development of the project. This process begins with the background or preliminary research phase, followed by the stage of defining the essential requirements for the construction of the test suite. This is followed by the development and containerization phase of the test suite. Next, the execution strategy to be employed is established, culminating with the obtaining of results that allow us to answer the initially formulated research questions.

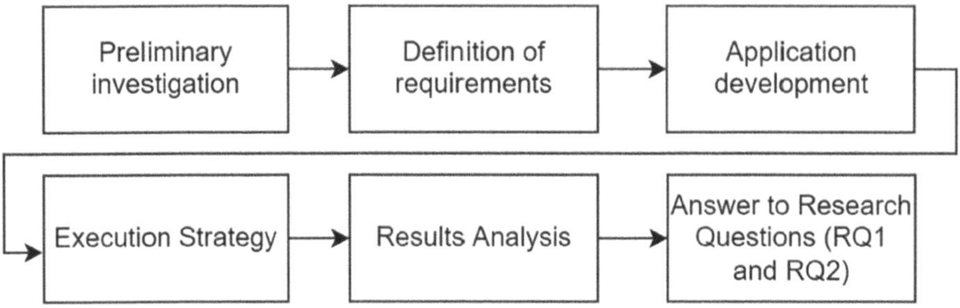

Fig. 1. Project methodology

3.1 Test Suite Design

The test suite includes 12 test cases designed to evaluate both local and remote accesses. These are combined with CPU stress tests, cache coherence, and verification of the cores allocated in each NUMA, to ensure the correct operation of the NUMA balancing feature. Such evaluation is performed both in the container environment and within the VM. It is important to note that, out of the 12 initial test cases, only 7 were selected through pairwise testing techniques, specifically through the AllPairs technique. This was done to optimize the execution while maintaining the same coverage. Verification was performed on each of the NUMA clusters tested, resulting in a total of seven cases instead of six. The steps that were carried out for pairwise testing are shown in Fig. 2 [22,23].

The test cases selected using the pairwise test technique are as follows in Table 1:

Table 1. List of test cases using pairwise testing

Test Case	Description
Test Case 1	Bare metal execution -Cache stress - SNC2
Test Case 2	Virtualization execution - Remote stress - SNC2
Test Case 3	Virtualization execution - Local stress - SNC2
Test Case 4	Bare metal execution - Remote stress - SNC4
Test Case 5	Bare metal execution - Local stress - SNC4
Test Case 6	Virtualization execution - Cache stress - SNC4
Test Case 7	Virtualization execution - Local stress - SNC4

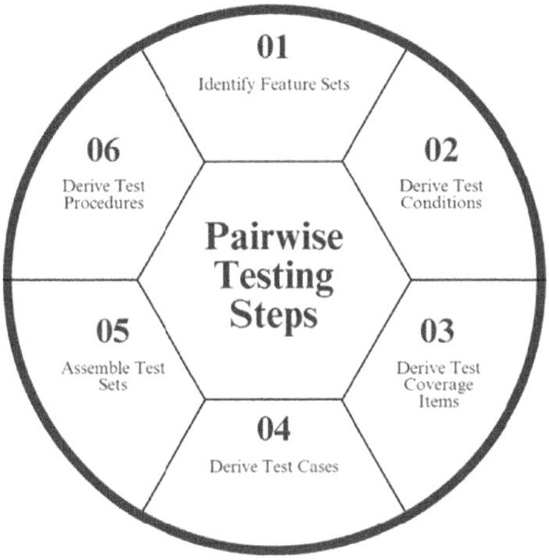

Fig. 2. Pairwise testing steps

Importantly, the psutil library proved crucial for efficiently obtaining information about system resources, such as system memory, total core count, and the number of NUMA nodes [24].

In addition, it is relevant to note that for the test cases focused on virtualization, a VM was configured using QEMU-KVM [25]. In this configuration, 4 NUMA virtual nodes were added, reserving a portion of the total memory minus 40 MB, which was divided by 4 to allocate it to the Memory-backend-ram object. A similar method was used to allocate cores; 12 were subtracted from the total number of available cores in the multisocket system and the result was divided by 4 to distribute them among each node, assigning each one its corresponding node ID.

3.2 Implementation of the Suite

The implementation and development of the test suite was carried out using Python, through which five fundamental scripts were created: main, platforminfo, verification, stress, and virtualization. Initially, the main script incorporates a small command line interface that facilitates the dynamic use of the tool. This script also organizes the execution flow of the test suite, grouping each test into an array for sequential execution.

The platforminfo script is where psutil deploys its functionality, gathering vital system information such as the amount of memory, the number of cores (core count), and the number and distribution of NUMA nodes. This data is essential for the effective execution of the remaining three scripts.

The verification script is responsible for confirming that the NUMA balancing has been performed correctly. It performs a comparison between the current distribution of system cores among the NUMA nodes and the expected distribution. If the values match, the process continues; otherwise, the execution is stopped and marked as "TEST FAILED".

The stress script contains the functions for the test cases that induce stress on the system. This includes everything from local and remote accesses, to CPU tests and cache coherence.

Finally, the virtualization script takes care of the creation and destruction of VMs, thus allowing stress tests to be run within them. This modular and detailed approach ensures a thorough and accurate evaluation of the system under test.

Docker was used to create the container. Using a Dockerfile, we proceeded to install and add multiple packages, dependencies, and files necessary for the operation of the test suite. Initially, the CentOS 8 operating system was installed, followed by the addition of various packages. "These included stressors such as stress-ng, stressapptest, virtualization tools such as qemu-kvm, programming languages such as Python, and utilities such as git and sshpass" [26,27]. In addition, several Python libraries were installed, including argparse, regex, and psutil. Stressapptest was also added using git.

As an image for the VM, we used an Ubuntu Bionic image that was provisioned with all the necessary packages before its execution by means of the Packer tool.

3.3 Test Strategy for Cluster Environments

In order to perform the experiments, it was essential to use Kubernetes, which requires a control plane equipped with critical elements such as API, etcd, scheduler, and control manager to manage the nodes based on execution. In addition, each of the systems within the cluster-called "nodes"-needed to incorporate three key components to perform the deployment: Kubelet Service, Kube-proxy, and Container Runtime Interface. The architecture of the cluster, from the control plane to the nodes to run the tests, is illustrated in Fig. 3.

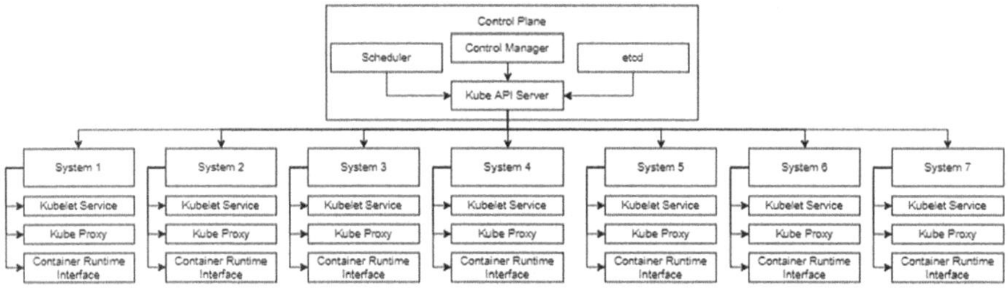

Fig. 3. Kubernetes cluster structure

However, to implement the execution strategy, it was planned that a selected set of tests would be carried out, each with a duration of 2 h, for a total of 14 h of execution. These tests were selected using the pairwise technique, with the objective of optimizing the coverage of the characteristics to be evaluated. In addition, they were performed on seven systems, distributed in two different experiments.

The execution flow, which commenced with the initial setup and proceeded through the NUMA clustering modification, followed by the test execution, and culminating in the verification via Kubernetes health checks, is illustrated in Fig. 4.

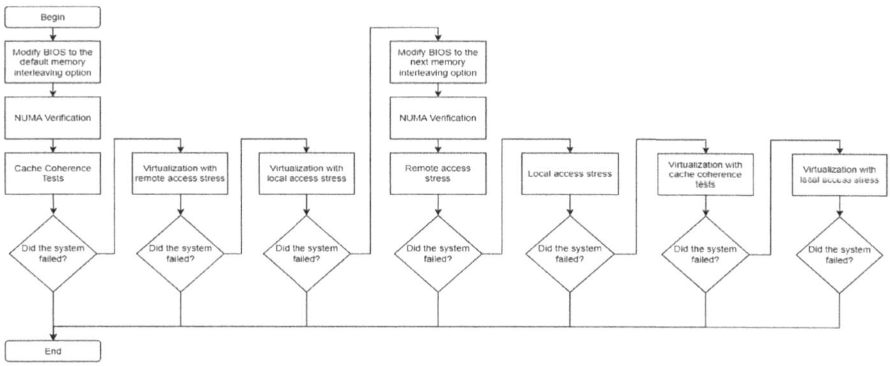

Fig. 4. Test suite execution flow

3.4 Interviews Based on Results

Finally, interviews were conducted with experts in system testing who had specialization in memory and processors. For this purpose, a sample of only 10 engineers, aged between 25 and 35 years, with more than 3 years of experience in the area was selected. The limited number of participants was attributed to the difficulty of finding individuals with these specific characteristics during the research period. During the interview, some questions were asked to learn more about what individuals thought of the test suite to ensure that from an expert's point of view, the test suite and the methodology can be effective and efficient. The questions were as follows:

1. What are the strategies and tools that you have found most effective for addressing non-uniform memory access in post-silicon validation stages?
2. How do you ensure consistency and accuracy in non-uniform memory access in post-silicon systems?
3. What recent trends or developments have you observed in the validation of non-uniform memory access in post-silicon systems?
4. Do you consider that the use of containers facilitates test portability and saves execution time? Why?

5. Do you think that the pairwise methodology can make the execution time for issue replication on servers more efficient?
6. Do you consider that deploying containers through Kubernetes has any advantage over other techniques used?

4 Results

In this section, we will present the results derived from the experiments conducted, highlighting the time efficiency achieved using cloud tools, as well as significant information obtained from the interviews.

Beginning with the quantitative results section, we present the data collected from both experiments. In the first experiment, five systems were evaluated, while in the second experiment, two systems were analyzed. The specific details of each experiment are available for consultation in the attached Table 2.

Table 2. Experiment results

Last Information				
System	Sub Numa Cluster	Point of Failure	Error	Test Result
System 1	Default, Numa per Socket 2	NA	No Error Detected	Test Passed
System 2	Default	Test Case 1	Uncorrectable Memory Error	Test Passed
System 3	Default, Numa per Socket 2	NA	No Error Detected	Test Passed
System 4	Numa per Socket 2	Test Case 7	Uncorrectable Memory Error	Test Passed
System 5	Default, Numa per Socket 2	NA	No Error Detected	Test Passed
System 6	Numa per Socket 2	NA	No Error Detected	Test Passed
System 7	Default, Numa per Socket 2	NA	No Error Detected	Test Passed

Among the results obtained from the first experiment, which involved the execution of five systems, it was observed that two of them experienced uncorrectable memory errors, evidencing failures in the memory banks. This indicates that 40% of the systems evaluated in this experiment presented failures. These failures were identified by the system health check provided by Kubernetes, revealing that the affected nodes were in a 'Not Ready' state. This suggests the possible loss of one or more of the three components essential to the operation of the system: the Kubelet service, kube proxy, or the container runtime interface. Upon further investigation, it was discovered that the affected systems were found to be unresponsive and in a failed state. The detection of uncorrectable memory errors was observed through Serial over LAN where the system failure data could be observed.

Subsequently, in the second experiment, both systems survived the tests. Therefore, it can be said that the execution was clean. Also, it should be mentioned that the memory correctable errors through error correcting code (ECC) were not considered for the results of this work. Additionally, a comparative analysis was carried out between the time required to run containers manually and the execution time using Kubernetes, with the objective of highlighting the efficiency that these technologies provide. The details of this comparison are presented in the following Table 3.

Table 3. Execution time manual vs Kubernetes

Execution Type	Time (Average)
Manual execution	1 min 37 s/per system
Kubernetes execution	2 min 52 s/all systems

The result in terms of efficiency using Kubernetes is a reduction of 77% of the time considering that the 7 systems were launched in a single deployment and that the container did not have to be re-executed after changing the knobs in BIOS. In the case of two or more nodes per deployment, it turns out to be more efficient.

On the other hand, we have the efficiency within the test cases reduced by means of the pairwise technique that can be visualized in Table 4.

Table 4. Time execution between tests

Execution Type	Comprehensive Test Cases/Time	Pairwise Test Cases/Time
Default NUMA Clustering	6 Test Cases/12 h	3 Test Cases/6 h
2 NUMA Node per Socket	6 Test Cases/12 h	4 Test Cases/8 h
Total	12 Test Cases/24 h	7 Test Cases/14 h

The results obtained from the interviews reveal that while some questions generated uniform responses, others elicited a wide variety of responses. From the findings obtained through this research tool, the conclusions listed below can be deduced. Also, it should be noted that in this section the qualitative part is only tangentially observed.

- P2: "The current tendency to find issues is through stress and injection error."
- P3: "Based on the results obtained and the manual testing done in the past, cloud technologies definitely automate and reduce the execution time."
- P5: "I believe that based on the experiment, containers can generate efficiency in the execution processes in post-silicon stages."
- P5: "Kubernetes can also be a good way to run tests in parallel. However, I don't know the implications that may affect the testing environment."

In conclusion, by highlighting the main benefits of the test suite, as well as the use of cloud technologies and the results obtained from the experiments, it is possible to clearly determine the following benefits:

- Emulation can be obtained through NUMA nodes to generate a CSP-like environment for testing.
- Portability of the suite with tools such as Docker containers and provisioned images for VMs.
- Parallel system execution and system health check capabilities with Kubernetes.

5 Discussion

In approaching the analysis and interpretation of the results obtained in this study, we are faced with the challenge of answering the research questions initially posed. In this discussion segment, we will answer in detail the research questions and the implications they have on the field of study.

The following is a comprehensive analysis of the results obtained:

5.1 RQ1: NUMA Suite Effectiveness

A. RQ1: NUMA Suite Effectiveness To answer the research question about the effectiveness of the test suite, it can be stated that after performing two experiments in which seven systems were run, two of them exhibited very similar errors related to uncorrectable failures in memory banks. This indicates that 29% of the systems evaluated in the experiment manifested this type of behavior.

Furthermore, it is important to note that 100% of the systems were not expected to exhibit failures, since, in theory, the systems should be stable, and it should be difficult to induce this type of error.

5.2 RQ2: Cloud Tools Efficiency

When addressing the question about the efficiency of tools commonly used in the cloud, such as Docker and Kubernetes, the following conclusion is reached:

The portability offered by containers is extremely efficient compared to traditional manual execution methods, which involve installing multiple packages individually. While this process is not too time consuming, using Docker containers allows you to have everything pre-installed, which saves the minutes spent installing Linux packages or Python libraries during manual testing. These time savings, while may seem minor in individual instances, translate into significant hours per year when applied to hundreds of systems or multiple runs.

Kubernetes as an execution tool can truly save time provided that two or more nodes are run, which was the case for both experiments. However, Kubernetes offers more than just that; it allows for the monitoring of events through it, from CPU usage, memory, and related metrics via its API. This tool provides a world of possibilities for such projects.

Finally, based on the interviews, some of the system validation engineers consider empirically that the use of these tools can contribute positively to the validation of this subsystem in servers.

6 Conclusions and Future Work

In conclusion, the feasibility of developing a test suite using open-source tools such as containers, orchestrators and stress generation tools for memory and processor fault detection has been demonstrated. The test suite is capable of exercising both local and remote accesses on servers with shared memory architecture where two uncorrectable memory errors were found to lead systems to failure, especially for NUMAs across a cluster environment. This represents a breakthrough in verifying the stability of CSP systems.

Looking ahead, it is possible to run the test suite on dozens of systems to verify its effectiveness on heterogeneous clusters, which present diverse characteristics in terms of memory, processors, and other hardware components. This strategy will provide a deeper understanding of the suite's ability to identify errors based on the diversity of hardware configurations.

On the other hand, it can be added that within the diagnostics of NUMA-based systems, there is a vast opportunity for expansion. Promising areas are identified for the development of new functionalities to be tested, including NUMA variants such as cache coherent non-uniform memory access (ccNUMA), non-remote memory access (NORMA), cache-only memory architecture (COMA), distributed shared memory (DSM), and shared virtual memory (SVM). These variants present opportunities for fault detection in more complex systems.

The integration of specific features for Compute Xpress Link (CXL) could also significantly improve fault detection in remote access environments [28], as it facilitates direct access from I/O to memory and adds an additional NUMA node to the system.

Finally, the incorporation of additional open-source tools to generate system stress and perform benchmarking tests will allow a more comprehensive evaluation of performance in multiple memory interleaving configurations. Increased coverage in this aspect can provide a solid basis for improvement in future research, experiments, and developments in this field.

Acknowledgement. I would like to express my gratitude to the people who contributed directly or indirectly to this work. Special thanks to my family and friends for their unconditional support. This project would not have been possible without their help.

References

1. Lameter, C.: NUMA (Non-Uniform Memory Access): an overview: NUMA becomes more common because memory controllers get close to execution units on microprocessors. Queue **11**(7), 40–51 (2013). https://doi.org/10.1145/2508834.2513149
2. Pan, X., Mueller, F.: NUMA-aware memory coloring for multicore real-time systems. J. Syst. Archit. **118**, 102188 (2021). https://doi.org/10.1016/j.sysarc.2021.102188

3. Ploskas, N., Samaras, N.: GPU Programming in MATLAB, pp. 6–7. Elsevier Inc. (2016). ISBN 978-0-12-805132-0
4. Lameter, C.: Local and Remote Memory: Memory in a Linux/NUMA System. Silicon Graphics, Inc. (2006). https://www.ilinuxkernel.com/files/Local.and.Remote.Memory.Memory.in.a.Linux.NUMA.System.pdf
5. Majo, Z., Gross, T.R.: (Mis)understanding the NUMA memory system performance of multithreaded workloads. In: 2013 IEEE International Symposium on Workload Characterization (IISWC), pp. 11–22 (2013). https://doi.org/10.1109/IISWC.2013.6704666
6. Popov, M., Jimborean, A., Black-Schaffer, D.: Efficient thread/page/parallelism autotuning for NUMA systems. In: Proceedings of the ACM International Conference on Supercomputing, ICS 2019, pp. 342–353 (2019). https://doi.org/10.1145/3330345.3330376
7. Guo, X., Han, H.: The research of a memory accesses behavior on non-uniform memory access architecture. In: Proceedings of the 10th International Conference on Information Technology in Medicine and Education (ITME), pp. 763–766 (2019). https://doi.org/10.1109/ITME.2019.00174
8. Dell Technologies: Memory population rules for 3rd generation Intel Xeon Scalable processors on PowerEdge servers (2023). https://infohub.delltechnologies.com/es-es/t/memory-population-rules-for-3rd-generation-intel-xeon-scalable-processors-on-poweredge-servers-1/
9. Advanced Micro Devices: Socket SP3 platform NUMA topology for AMD Family 19h Models 00h-0Fh. Publication #56795, Revision: 1.00 (2021)
10. Vladimirov, A., Asai, R.: Clustering modes in Knights Landing processors: Developer's guide. Colfax International (2016). https://colfaxresearch.com/knl-numa/
11. Liu, D., Zhang, Y. Y., Zhang, N., Hu, K.: A research on KVM-based virtualization security. In: Vehicle, Mechatronics and Information Technologies II, Applied Mechanics and Materials, vol. 543, pp. 3126–3129. Trans Tech Publications Ltd. (2014). https://doi.org/10.4028/www.scientific.net/AMM.543-547.3126
12. Docker Architecture. https://docs.docker.com/get-started/overview/. Accessed 21 July 2024
13. Pahl, C.: Containerization and the PaaS cloud. IEEE Cloud Comput. **2**(3), 24–31 (2015). https://doi.org/10.1109/MCC.2015.51
14. Preeth, E.N., Mulerickal, F.J.P., Paul, B., Sastri, Y.: Evaluation of Docker containers based on hardware utilization. In: Proceedings of the 2015 International Conference on Control Communication & Computing India (ICCC), pp. 697–700 (2015). https://doi.org/10.1109/ICCC.2015.7432984
15. Poulton, N.: The Kubernetes Book. Leanpub (2021). ISBN 978-0-9814-3712-4. http://leanpub.com/thekubernetesbook
16. Todorov, M.H.: Deploying different lightweight Kubernetes on Raspberry Pi cluster. In: 30th National Conference with International Participation (TELECOM), pp. 1–4. IEEE (2022). https://doi.org/10.1109/TELECOM56127.2022.10017262
17. Todorov, M. H.: Design and deployment of kubernetes cluster on raspberry pi OS. In: 29th National Conference with International Participation (TELECOM), pp. 104–107. IEEE (2021). https://doi.org/10.1109/TELECOM53156.2021.9659651
18. Monteiro, C.B.A.L., Dias, L.A.V., da Cunha, A.M.: A case study on pairwise testing application. In: 11th International Conference on Information Technology: New Generations, pp. 639–640. IEEE (2014). https://doi.org/10.1109/ITNG.2014.42
19. Ismail, H. A., Riasetiawan, M.: CPU and memory performance analysis on dynamic and dedicated resource allocation using XenServer in Data Center environment.

In: 2nd International Conference on Science and Technology-Computer (ICST), pp. 17–22. IEEE (2016). https://doi.org/10.1109/ICSTC.2016.7877341
20. Stepanovic, S., Georgakarakos, G., Holmbacka, S., Lilius, J.: Quantifying the interaction between structural properties of software and hardware in the ARM Big.LITTLE architecture. In: 26th Euromicro International Conference on Parallel, Distributed and Network-based Processing (PDP), pp. 138–144. IEEE (2018). https://doi.org/10.1109/PDP2018.2018.00027
21. Dixit, D., Lin, F. (Fred), Holland, B., Beadon, M., Yang, Z., Sankar, S.: Optimizing interrupt handling performance for memory failures in large scale data centers. In: Proceedings of the ACM/SPEC International Conference on Performance Engineering, pp. 193–201. ACM (2020). https://doi.org/10.1145/3358960.3375793
22. IEEE: ISO/IEC/IEEE International Standard - Software and systems engineering – Software testing – Part 1: General concepts. In: ISO/IEC/IEEE 29119-1:2022(E), pp. 1–60. IEEE (2022)
23. IEEE: IEEE Approved Draft International Standard for Software and Systems Engineering–Software Testing–Part 4: Test Techniques. In: ISO/IEC/IEEE P29119-4-FDIS, pp. 1–147. IEEE (2015)
24. Psutil Python. https://pypi.org/project/psutil. Accessed 21 July 2024
25. About QEMU. https://ww.qemu.org/docs/master/about/index. Accessed 21 July 2024
26. Stress-ng. https://wiki.ubuntu.com/Kernel/Reference/stress-ng. Accessed 21 July 2024
27. Stressapptest. https://github.com/stressapptest/stressapptest. Accessed 21 July 2024
28. Sun, Y., et al.: Demystifying CXL memory with genuine CXL-ready systems and devices. In: Proceedings of the 56th Annual IEEE/ACM International Symposium on Microarchitecture, pp. 105–121. IEEE/ACM, Toronto (2023). https://doi.org/10.1145/3613424.3614256

Information Technology

Countering Generative AI-Driven Social Engineering and Phishing: A Threat Modeling and Cybersecurity Awareness Framework

Miguel Prada(✉)

New York University, Wunsch Building 311 Bridge St Brooklyn, New York, NY 11201, USA
map10127@nyu.edu

Abstract. This paper explores the growing threat of GenAI in social engineering and phishing attacks. We analyze how GenAI can be misused, and the potential consequences of this. The study describes the impact of threat modeling and cybersecurity awareness training as safeguards. By identifying vulnerabilities and empowering users, we aim to develop a roadmap for mitigating these emerging GenAI-driven attacks.

Keywords: Phishing · social engineering · generative artificial intelligence · threat modeling · cybersecurity awareness training

1 Introduction

Generative Artificial Intelligence (GenAI) constitutes a powerful emerging technology with the potential to be misused for malicious purposes. Large language models (LLM) such as ChatGPT, Microsoft Copilot, and Gemini (formerly Google Bard), are a specific type of GenAI that focus on language-based tasks. With the increasing sophistication of LLMs, their potential for harm in social engineering and phishing attacks grows. These attacks can result in significant financial losses, data breaches, and reputational damage [1].

This research proposes the combination of comprehensive threat modeling with cybersecurity awareness training, as a novel two-pronged approach that can be deployed to combat cybersecurity threats using various delivery methods, to raise end-user awareness and foster secure behavior. The overall objectives of the research were to empower users to recognize and report malicious activities in a timely manner, and to encourage best cybersecurity practices to be used in daily routines.

Threat modeling is a process to systematically identify and assess potential vulnerabilities, and different case studies are discussed in this work to explore the risks arising from GenAI capabilities. Additionally, cybersecurity awareness training is crucial to ensure that end users can make appropriate choices to

F. Suni-Lopez et al. (Eds.): CIIS 2024, ISEM 48, pp. 181–189, 2025.
https://doi.org/10.1007/978-3-031-91481-2_12

prevent or mitigate the risks posed by social engineering and phishing attacks. User education campaigns that are tailored based on the identified threats will equip users with the knowledge to recognize and resist these attacks [2,3].

2 Methodology

This paper is structured as follows: Sect. 2 will discuss the research methodology used in this work, including a motivating case study which is reviewed and discussed using threat modeling; in Sect. 3, conclusions and future work is discussed.

A growing body of research suggests that cybersecurity awareness programs can mitigate human factors contributing to cybersecurity vulnerabilities, particularly those related to social engineering and phishing attacks [2,3]. For instance, the study by Sheng et al. cited in [3] reported a 40% reduction in users entering sensitive information on fraudulent websites after receiving cybersecurity awareness training. These findings highlight the critical role of cybersecurity awareness training in enhancing organizational resilience against social engineering and phishing attempts. A diverse range of intervention programs are available, employing training, education, and persuasive techniques to equip end-users with the necessary skills to identify and avoid phishing attacks [3]. As presented in Table 1, there are different authors conducting research related to cybersecurity awareness training as a countermeasure for GenAI driven cyber-attacks.

Phishing attacks have long plagued the digital landscape, and previously, such attacks relied on generic templates and social engineering tactics that preyed on human vulnerabilities [2,6]. As presented in Table 2, three security domains are referenced in the literature for GenAI driven cyber-attacks.

Threat modeling helps identify, assess, and mitigate potential threats by systematically analyzing attack scenarios [8]. Many threat modeling approaches have been reported ranging from textual descriptions to graphical methods [8,9]. Figure 1 presents the threat modeling process applied to this work.

A threat modeling analysis should start with a well-defined scope and asset identification. As a case study, the author has considered an e-commerce chatbot powered by LLM. Following the process referenced in Fig. 1, the assets requiring protection include user data (names, contact information, payment details, purchase history), shopping cart information (items, quantities, prices), chat transcripts, chatbot training data, and system access credentials. For this motivational example we have applied the STRIDE method for attack vector identification (see Table 3). The STRIDE method refers to six domains: spoofing, tampering, repudiation, information disclosure, denial of service, and elevation of privilege [10].

This study posits that a combination of threat modeling to understand GenAI's deceptive content capabilities and cybersecurity awareness training on such tactics can significantly reduce attack success rates. For this work, relevant available empirical evidence was used (see Table 4). As per data reviewed in the literature, quantitative methods are preferred for assessing security vulnerabilities arising from social engineering attacks. Authors in [11–16] examined this

Table 1. Suggested approach based on related research found about cyber awareness training as a countermeasure for GenAI driven attacks

Author	Conclusion found in the literature	Approach suggested by the author
[4]	Studies show that educating online users about phishing and implementing proper anti-phishing measures can significantly reduce the risk of falling victim to these attacks. This two-pronged approach is crucial for protecting online identities from phishing scams	A continual improvement of cybersecurity awareness programs by challenging online consumers to learn and put in practice the threat modeling methods through educational anti-phishing activities
[2]	AI is fundamentally changing security across all domains-digital, physical, and political. It strengthens both attack and defense capabilities in the cyber realm, while creating new vulnerabilities for hackers. While the specific risks of malicious AI use are vast, understanding these commonalities across different security areas is crucial to prepare for future threats and improve prevention strategies.	LLMs can improve user experience and cybersecurity training materials with case studies, risk identification, queries, and risk assessment capabilities, if proactive threat modeling can be integrated to GenAI embedded in enterprise SaaS applications
[5]	Despite limitations, user awareness training can enhance preparedness against AI-powered cyberattacks. By fostering knowledge of evolving cybercriminal tactics, such training empowers individuals to identify and respond effectively to potential threats. This can be achieved through well-designed training programs, interactive simulations, and accessible educational materials.	Train employees with personalized phishing simulations. These emails, powered by LLMs, would mimic real-world attacks by considering individual work habits, interests, and online behavior. This approach reflects how attackers personalize scams, making the training more effective
[6]	Traditionally, social engineering attacks are tackled with security rules and staff training. However, research suggests these methods need improvement. Gragg (2003) emphasizes including psychology and persuasion techniques in training to better understand and defend against these attacks. Similarly, Srikwan (2008) proposes using cartoons to educate users about social engineering and phishing tactics	Build a game-like platform where users practice spotting security risks. This platform would use AI to simulate real attacks, letting users make choices and learn from their decisions to strengthen their security awareness
[7]	Effective design, implementation, and evaluation of anti-phishing interventions are key to enabling organizations to turn their employees from a potential source of cyber security vulnerability into their strongest line of defense, by providing employees with the skills to identify and report phishing attacks	User-friendly reporting platforms with machine learning capabilities can be implemented to analyze suspicious emails, identify phishing tactics, and build a departmentalized knowledge base, ultimately enhancing user recognition of future threats

approach. User interaction and information processing are identified as areas susceptible to these attacks [12,13,17]. Other studies like Sheng et al. cited in [3] reported a 40% reduction in users entering sensitive information on fraudulent websites after receiving cybersecurity awareness training. These findings

Table 2. GenAI driven cyberattacks and their security domains

Security Domains	Description	Example
Digital Security	The GenAI automation can significantly amplify the threat associated with labor-intensive cyberattacks, by enabling the generation of highly targeted and personalized content at scale.	GenAI can facilitate the development of novel attack vectors, exploiting human vulnerabilities, existing software vulnerabilities, or even the vulnerabilities of AI systems them-selves
Physical Security	GenAI can outline attacks with drones and other physical systems	As a combined method, this work suggests that LLMs can improve user experience and cybersecurity training materials with case studies and risk identification
Political Security	To innovate in tasks involved in subverting cyber-physical systems, GenAI can reduce time required for surveillance, persuasion, and deception automation	Surveillance: Analyzing mass-collected data. Persuasion: Creating targeted propaganda. Deception: Manipulating videos

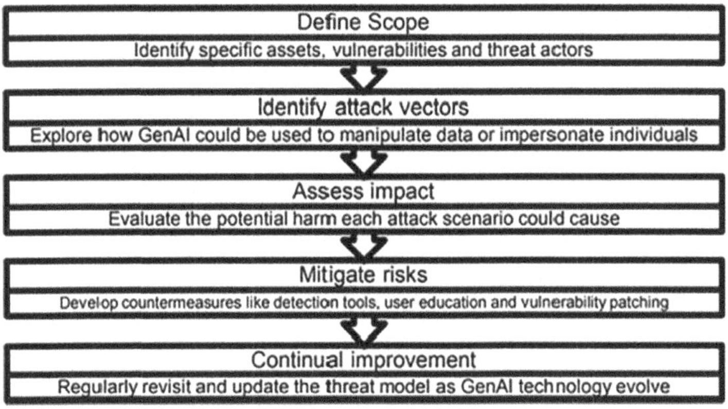

Fig. 1. Generic threat modeling process followed in this work

highlight the critical role of cybersecurity awareness training in enhancing organizational resilience against social engineering and phishing attempts. This highlights the importance of user-centric security frameworks to mitigate risks, as emphasized by various researchers [13, 14, 16].

Table 3. Threat modeling applied to the motivational example

Attack Vectors	STRIDE Domains	Assess Impact	Mitigate Risks
Exploiting user trust, bypassing security controls, leveraging data vulnerabilities, manipulating user behavior, abusing chatbot functionalities	Spoofing	Attackers could spoof the identity of legitimate representatives to extract data	Implement multi-factor authentication and clear warnings within the chatbot interface to inform users they are interacting with a chatbot, not a human representative
	Tampering	Attackers might manipulate chatbot responses through training data/code tampering	Secure training data, code, and monitor chatbot behavior for anomalies through access controls, audits, and monitoring
A data breach attack tree is presented in Fig. 2	Repudiation	Chatbot interactions involving misleading information may lead to financial disputes	Maintain detailed logs of user interactions with the chatbot, including timestamps, transcripts, and user choices made during the conversation
	Information Disclosure	Malicious actors could exploit vulnerabilities in the system to gain unauthorized access to user data.	Implement least privilege for chatbot access, minimizing data collection and storage. Encrypt sensitive data and conduct regular security audits to identify and fix vulnerabilities
	Denial of Service	Denial-of-service (DoS) attacks can disable the chatbot for real users	Implement an intrusion detection and prevention system (IDS/IPS) for enhanced security and to handle peak user loads
A functional disruption attack tree is presented in Fig. 3	Elevation of Privilege	Privilege escalation can compromise data or disrupt operations.	Least privilege access control and regular vulnerability patching safeguard the chatbot and infrastructure

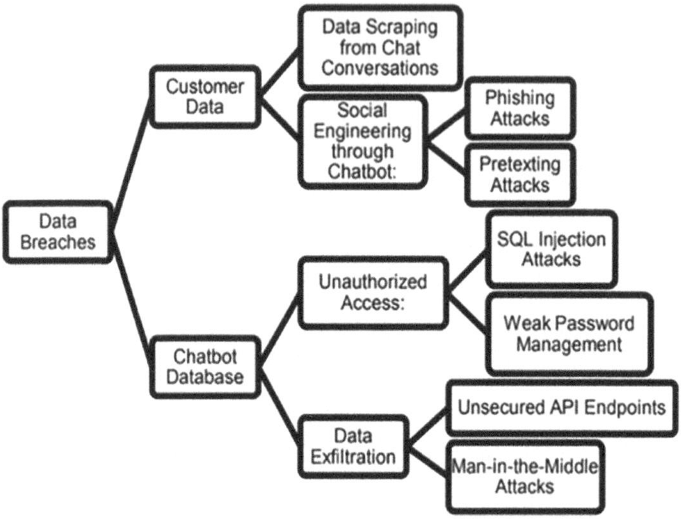

Fig. 2. Data Breach Attack Tree

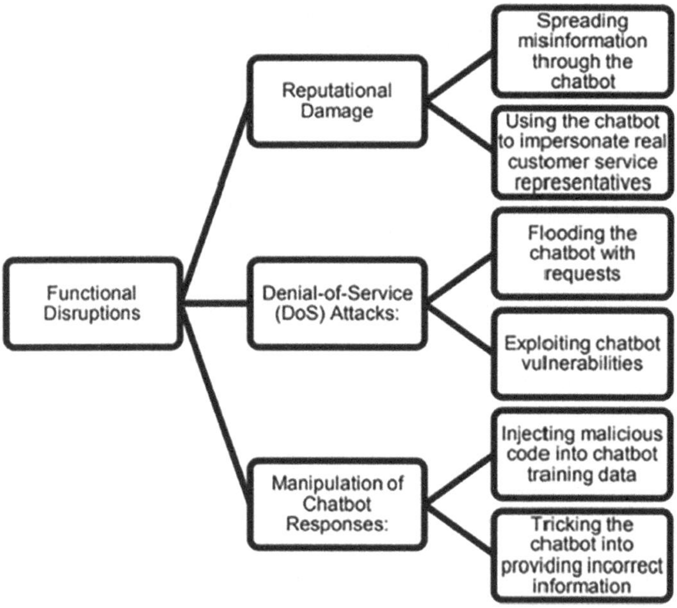

Fig. 3. Functional Disruption Attack Tree

Table 4. Empirical data found in the literature review

Author	Research Approach	Research Type	Security Attacks
[11]	Co-utile disclosure	Quantitative research	Information exchange
[12]	Personality information	Processing Model	Information processing
[13]	User-centric, preventing social engineering attacks	Quantitative research	Social engineering attacks
[14]	Human as a security sensor framework	Quantitative research	Social engineering attacks
[15]	Security and privacy in an online social networking model	Quantitative research	Security and privacy
[16]	User vulnerability to the social engineering model	Quantitative research	Social engineering attacks
[17]	User interaction framework	Qualitative research	User interaction

Table 5. Methodology suggested for future work

Steps	Expected Outcome
Sample Selection	Random sampling of employees from a chosen industry. Stratified sampling to ensure representation of different departments and experience levels
Data Collection	Baseline survey: Assess initial knowledge and attitudes towards social engineering and phishing threats. User interviews/focus groups: Understand user experiences with social engineering attempts and identify common vulnerabilities. Pre-test: Conduct social engineering simulations to measure baseline susceptibility. Implement user-centric training: Develop training programs based on user experiences, vulnerabilities, threat modeling case studies, and preferred learning styles. Post-test: Conduct social engineering simulations after training to measure improvement
Data Analysis	The study's efficacy will be assessed through pre- and post-test analyses. Social engineering attempt reduction will be quantified, and user knowledge/attitudes will be compared pre- and post-training. Statistical tests (e.g., t-tests) will determine the significance of these changes, evaluating the user-centric approach's effectiveness
Conclusions	Demonstrate a decrease in successful attacks after training. Show an improvement in user knowledge and awareness of social engineering tactics. Highlight the importance of tailoring cyber security awareness programs to user needs and perspectives

3 Conclusions and Future Work

This paper presents a user-centric approach to evaluating cybersecurity awareness programs against social engineering threats. The proposed methodology integrates threat modeling principles into cybersecurity awareness training, enabling users to identify attack vectors and social engineering tactics. This data-driven framework empowers users to become active participants in their own online security (see Table 5). Educating online users about phishing and implementing proper anti-phishing measures can significantly reduce the risk of falling victim to these attacks. The suggested approach of the author emphasizes continual improvement of cybersecurity awareness programs, by challenging users to learn and put into practice threat modeling methods, through educational anti-phishing activities.

This work also suggests that LLMs can improve training materials with case studies, risk identification, and risk assessment following a threat modeling perspective in cybersecurity awareness training. Also personalized phishing simulations, powered by LLMs, would mimic real-world attacks by considering individual work habits, interests, and online behavior. By personalizing phishing simulations, organizations can create a more realistic learning environment that better prepares employees to identify and respond to actual phishing attempts. User-friendly reporting platforms with machine learning capabilities can be implemented to analyze suspicious emails, identify phishing tactics, and build a departmentalized knowledge base, ultimately enhancing user recognition of future threats.

References

1. Crothers, E.N., Japkowicz, N., Viktor, H.L.: Machine-generated text: a comprehensive survey of threat models and detection methods. IEEE Access **11**, 70977–71002 (2023). https://doi.org/10.1109/ACCESS.2023.3294090
2. Brundage, M., Avin, S., Clark, J., Toner, H.: The malicious use of artificial intelligence: forecasting, prevention, and mitigation. arXiv preprint arXiv:1802.07228 (2024). https://doi.org/10.48550/arXiv.1802.07228
3. Alhashmi, A.A., Darem, A., Abawajy, J.H.: Taxonomy of cybersecurity awareness delivery methods: a countermeasure for phishing threats. Int. J. Adv. Comput. Sci. Appl. **12**(10) (2021). https://doi.org/10.14569/IJACSA.2021.0121004. https://www.proquest.com/scholarly-journals/taxonomy-cybersecurity-awareness-delivery-methods/docview/2655113201/se-2
4. Purkait, S.: Phishing counter measures and their effectiveness - literature review. Inf. Manag. Comput. Secur. **20**(5), 382–420 (2012)
5. Schmitt, M., Flechais, I.: Digital deception: generative artificial intelligence in social engineering and phishing. Artif. Intell. Rev. **57**(12), 324 (2024)
6. Krombholz, K., Hobel, H., Huber, M., Weippl, E.: Advanced social engineering attacks. J. Inf. Secur. Appl. **22**, 113–122 (2015)
7. Sarker, O., Jayatilaka, A., Haggag, S., Liu, C., Babar, M.A.: A multi-vocal literature review on challenges and critical success factors of phishing education, training and awareness. J. Syst. Softw. **208**, 111,899 (2024). https://doi.org/10.1016/j.jss.2023.111899. https://www.sciencedirect.com/science/article/pii/S0164121223002947
8. Hussain, S., Kamal, A., Ahmad, S., Rasool, G., Iqbal, S.: Threat modelling methodologies: a survey. Sci. Int. (Lahore) **26**(4), 1607–1609 (2014). https://www.researchgate.net/publication/307902746_THREAT_MODELLING_METHODOLOGIES_A_SURVEY
9. Falade, P.V.: Decoding the threat landscape: chatgpt, fraudgpt, and wormgpt in social engineering attacks. Int. J. Sci. Res. Comput. Sci. Eng. Inf. Technol. (IJSRCSEIT) **9**(5), 185–198 (2023)
10. Mauri, L., Damiani, E.: Modeling threats to ai-ml systems using stride. Sensors **22**(17) (2022). https://doi.org/10.3390/s22176662. https://www.mdpi.com/1424-8220/22/17/6662

11. Sánchez, D., Domingo-Ferrer, J., Martínez, S.: Co-utile disclosure of private data in social networks. Inf. Sci. **441**, 50–65 (2018). https://doi.org/10.1016/j.ins.2018.02.010
12. Frauenstein, E.D., Flowerday, S.: Susceptibility to phishing on social network sites: a personality information processing model. Comput. Secur. **94**, 101,862 (2020). https://doi.org/10.1016/j.cose.2020.101862. https://www.sciencedirect.com/science/article/pii/S0167404820301346
13. Albladi, S.M., Weir, G.: User characteristics that influence judgment of social engineering attacks in social networks. HCIS **8**(1), 1–24 (2018). https://doi.org/10.1186/s13673-018-0128-7
14. Yasin, A., Fatima, R., Liu, L., Yasin, A., Wang, J.: Contemplating social engineering studies and attack scenarios: a review study. Secur. Priv. **2**(4), e73 (2019)
15. Albladi, S.M., Weir, G.: Predicting individuals - vulnerability to social engineering in social networks. Cybersecurity **3**(1), 7 (2020)
16. Yusof, Y.M., Singh, D.: Civil servants awareness guideline towards computer security policy: a case study at the manpower department, ministry of human resources. Asia-Pac. J. Inf. Technol. Multimedia (Jurnal Teknologi Maklumat dan Multimedia Asia-Pasifik) **10**(8), 86–99 (2021)
17. Pitchan, M.A., Omar, S.Z., Ghazali, A.: Amalan keselamatan siber pengguna internet terhadap buli siber, pornografi, e-mel phishing dan pembelian dalam talian. Jurnal Komunikasi: Malay. J. Commun. **35**(3), 212–227 (2019)

Designing a Minimum Viable Blockchain Ecosystem for Forest Asset Management: A Case for Colombia

Fredy Alexander Muñoz Cano[1(✉)], Gina Paola Maestre Góngora[2], and Adriana Patricia Villegas Quiceno[1]

[1] Cooperative University of Colombia, Medellín 050015, Colombia
fredy.munozc@campusucc.edu.co, adriana.villegas@ucc.edu.co
[2] University of Antioquia. 17, Medellín 050010, Colombia
gina.maestre@udea.edu.co

Abstract. A minimum viable ecosystem (MVE) is an ecosystem that allows participants to meet a more complex future ecosystem with the least effort, considering three crucial aspects of blockchain adoption: breadth, scale, and standards, which is a key step for the advancement of blockchain in each sector. This paper presents the design of a blockchain MVE for forest asset management following the methodology proposed by the World Economic Forum based on five steps. The presented methodology is deployed for a case study related to the commercialization of climate assets describing the value proposition, participants, system model, governance, future steps, and the preliminary model of an IT architecture supporting the MVE.

Keywords: Blockchain · minimum viable ecosystem · IT architecture

1 Introduction

Traceability can be termed as the ability to track information to promote data access and transparency. Available centralized traceability solutions are exposed to many problems and challenges, such as data manipulation, single points of failure, etc., which is why blockchain-based traceability solutions can address the shortcomings of centralized traceability solutions [1].

Blockchain is an emerging technology with high implementation potential to improve traceability performance, by providing full traceability and transparency; also as a distributed accounting technology, it is gaining recognition mainly with applications in the financial sector, such as Bitcoin. It is a new paradigm of applying multiple emerging information technologies including asymmetric encryption, distributed networks, peer-to-peer transmission, smart contracts, and consensus mechanisms to create permanent, immutable, authoritative, and time-stamped transaction records, enabling it to establish mutual

F. Suni-Lopez et al. (Eds.): CIIS 2024, ISEM 48, pp. 190–202, 2025.
https://doi.org/10.1007/978-3-031-91481-2_13

trust at low cost in a competitive environment of distrust without third parties [2]. Currently, technology-based companies have begun to explore different models for managing an ecosystem, where the minimum viable eco-system (MVE) approach emerges, guided by the premise of "start small with an ecosystem that an organization can control, then build from there" [3].

The scope of a technology-based ecosystem is to extend the use of the system or technology, through the co-evolution and interconnection of its actors, to solve technical problems within an industry. In a technology-based ecosystem, it is the complementary technological innovations, based on the core technology, that drive the ecosystem's value creation [4]. An ecosystem as a structure-which considers ecosystems as configurations of activity defined by a value proposition-starts with a value proposition and seeks to identify the set of actors that need to interact for the proposition to become a reality [5].

This paper presents the background and use cases of blockchain in forest asset management and presents a conceptual framework for a minimum viable ecosystem. Then, a methodology for the design of these ecosystems is presented. Finally, an application case is presented following the proposed methodology designing an MVE supported by a value proposition, identification of participants, an ecosystem model, and its governance and future preparation.

2 Background

2.1 Blockchain and Forest Asset Management

Blockchain is a distributed, immutable electronic database that stores a record of all transactions that have occurred over a network. The technology is considered applicable to a wide range of industries, social service initiatives, and environmental management concerns including forest asset management as shown below.

Blockchain technology can also be used to manage forest fires in remote areas. Response time, a single point of failure, workload balance, and cost issues are problems to which the central management of forest stations and traditional optimization techniques are susceptible.

The study [6] proposes an integrated drone-based blockchain framework in which technology, drone controllers, firefighters, and managers are users or nodes in the network and have access to data, data exchange is fast, and obstacles related to spatial distance are minimized. The study concludes with a discussion of the challenges and opportunities of combining blockchain with other new technologies to manage wildfires in remote areas. The authors describe the opportunities for advocates of market ecologism, as well as key challenges for conservation movements. They assess four of the ways in which blockchain is being leveraged to address persistent failures related to (1) financing reducing emissions from deforestation and forest degradation (REDD) projects; (2) accurate tracking, reporting, and verification of forest carbon emissions; (3) equity in benefit sharing; and (4) securing the rights of indigenous and forest-dependent communities for Reducing Emissions from Deforestation and Forest Degradation

(REDD+). In addition, they consider environmental governance issues related to existing and proposed cryptocurrency projects, highlighting why applications of blockchain technology in conservation deserve the scrutiny of critical scholars.

In this way [7] , examine voluntary markets and the role that forest carbon offsets have played in them, highlighting the evolution of REDD and the barriers to its growth. They then discuss the potential of blockchain to address some of the problems identified in the literature and the developing experience of the forest industry. They conclude that the technology has the potential to improve verifiability, reduce transaction costs, and to a lesser extent, help resolve issues of additionality and permanence. However, to fully evaluate and maximize its potential, more needs to be learned from the emergence of blockchain in pilot projects.

Deforestation is the main cause of habitat loss for animals, and one of the main causes of deforestation is the commercial use of timber where it is difficult to determine whether certain varieties of trees have been conserved and not overexploited. The work of [8] proposes a method to track and observe the use of these trees using image processing and blockchain technology. First, the actual number of trees in a given area is determined using image processing algorithms. Then, smart contracts are created to store the details of the felled trees for insertion into the blockchain network. The user interface is designed to allow different stakeholders to interact seamlessly with the blockchain network.

2.2 Minimum Viable Ecosystem

Minimum viable ecosystem is a methodology that focuses on the actors and the development of an ecosystem, supported by information technologies, in order to design new products and services from a mainly organizational and strategic approach, based on the principles of collaboration and open innovation. One of the most notable applications of MVE is in blockchain and distributed ledger technology (DLT) use cases as shown by the work of [9–11]. The application of MVE in blockchain and DLT use cases facilitates the interaction and management of stakeholders to promote innovative services and the development of decentralized products.

An MVE defined by the father of this approach-Rod Adner-is "The smallest configuration of activities (and partners) that can create sufficient evidence of value creation to attract new partners. Since the goal of MVE is to attract partners, the key contribution of customers in the MVE phase is not to generate profits, but to create evidence to drive partner engagement" [12].

In this sense, the MVE is the ecosystem of least complexity that facilitates stakeholders and actors to take ownership and understand in an iterative and incremental way about the more complex future ecosystem with the least effort as it is being built. This includes learning about the interrelationships and interactions between ecosystem participants, such as customers, partners, and governments, and specifically, how value should be shared and generated as a whole [13].

This concept has been applied in a number of use cases. [9] create an artifact focused on the critical elements of a blockchain project to help conduct strategic thinking while establishing a precise project roadmap with the goal of achieving self-sufficiency in the form of an MVE. The framework was evaluated in three case studies: a startup, a digital casino, and an academic project. The framework functions as a guiding principle that broadly helps to answer the "why", "who", "what" and "when" of the project, but has limitations in answering the "how".

In the work by [10], a value plan is proposed where all the components of an MVE are clearly laid out and where the arrangement of the elements necessary to offer the value proposition, their position, and their relationships are presented. It concludes that the multiplicity of actors that make up the ecosystem creates new trade-offs that affect the nature of the strategy. This extends the consideration beyond the question of value creation and capture, to the question of value distribution in the broader ecosystem and to what extent it is treated as a matter of negotiation, or as a strategic lever in the construction of the ecosystem.

On the other hand, [11] identifies and describes the specific characteristics of ecosystems in a network of DLT actors, and makes comparisons of software, technological, digital, innovation, product, and service ecosystems, and assesses how they fit with the DLT ecosystem. They conclude that these ecosystems may be part or subset of the DLT ecosystem, but it is the enterprise ecosystem type that best fits the DLT ecosystem conceptualization. Finally, [14] corrects systemic issues prior to blockchain implementation and develops specific performance metrics for the agreed MVE and associated processes.

3 Methodology

When designing a blockchain ecosystem, it is essential to consider some steps and questions to ensure that an organization has a model that can deliver the expected benefits to the lead organization and other stakeholders. It should also be noted that appropriate design and dynamics of a blockchain ecosystem does not guarantee the success of the project. However, it is almost certain that a poorly planned ecosystem will be a cause of its failure. In fact, this is the most common reason for not obtaining the expected results [3]. According to [15], most companies use blockchain in the wrong way, in an effort to show executives and detractors how it works, when what they really need is a clear understanding of the business value proposition and where blockchain can contribute to their business.

Below are the fundamental steps and questions for the design of the ecosystem based on blockchain technology, following the proposal of the World Economic Forum [3,15] as shown in Fig. 1 and Table 1.

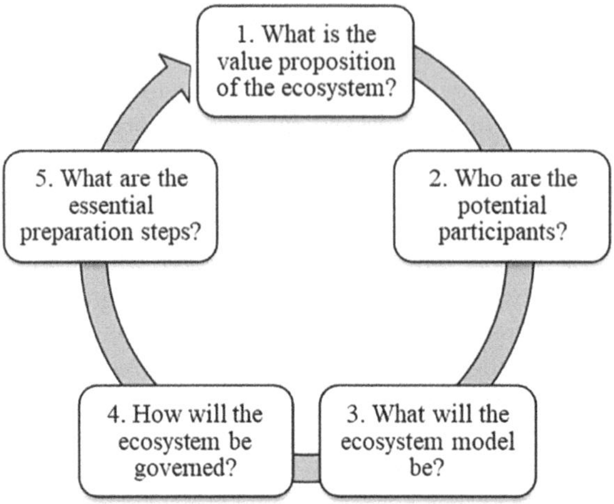

Fig. 1. MVE design methodology. Source: Based on [3]

Table 1. MVE design phases

Phase	Description
1. Ecosystem value proposition	Define why the organizations involved need a joint blockchain solution. Here you define the problem situation to be solved and the goal of the ecosystem
2. Participant's identification	In this step it is important to find out who wants and can invest in blockchain and will consider it essential technology. It is important to engage potential ecosystem stakeholders. Otherwise, it is possible to miss key design points that will encourage them to participate in the future
3. Ecosystem model	Here the organization will begin to get an idea of which collaboration model will best suit its objectives
4. Governance	Consideration should be given to who will manage the network and how it will be governed according to the characteristics of the participants and the models previously selected
5. Preparation of the organization	At this stage of the process, each organization should be clear about the benefits, opportunities, and limitations of joining the proposed ecosystem. However, before moving forward, a more detailed analysis process should be carried out

4 Case Study

In the Colombian context, the lack of controls and mechanisms to encourage the community not to intervene in the ecosystem for their economic benefit is evident, which is why the deforestation processes are growing year by year and the departments of Meta and Arauca in the eastern plains of Colombia are home to an area known as the Sarare nucleus where deforestation is particularly prominent, which is responsible for 0.1% of the national deforestation [16]. In the department of Meta, this problem was forecasted to affect 23,000 ha by 2019, and continues to grow, with the La Macarena Special Management Area (AMEM) being the most affected with nearly 10,471 ha of deforestation [16]. Other areas affected are those of preservation and recovery such as the integrated management districts of Ariari Guayabero in La Macarena and Uribe, Macarena Sur in La Macarena, and Macarena Norte in Puerto Rico and Vista Hermosa.

This situation means that carbon sequestration is decreasing because there are no agents which help CO2 to be captured in the soil in a natural way. This leads to the conclusion that there is insufficient incorporation of decarbonization actions for climate change mitigation, nor the protection of environmental assets in the department to strengthen conservation policies.

Although deforestation has been mitigated somewhat, this effort has been insufficient, because forests continue to be affected and it can be said that in the control process there is not enough human capital or technologies to help monitor ecosystem areas of special environmental interest, thus reaffirming the need to create new alternatives to mitigate deforestation. The survey presented in [16] states that "70% of those surveyed consider that the community in general is responsible for conserving the environment; 40% that it is the territorial entities; 30% the state or the government; and 20% others, related as follows: 10% believe that it is everyone's responsibility, and the other 10% the public sector, and the private entities.".

To this end, within the framework of the project "Implementation of actions for the protection of water and soil basins through reforestation with emerging technologies and biotechnology", a blockchain platform is proposed that involves the community, government, and companies. The purpose of the blockchain is to manage the primary information that serves as the basis for the calculation of the amount of carbon captured; the delimitation of ecosystem boundaries, size, and shape; and biodiversity. Current management and protection models for strategic environmental assets in the departments are either insufficient or entirely lacking. These models fail to serve as effective tools for decision-making. An improved approach should include mechanisms to assign economic value to these assets, enabling their monetization as a strategy to promote soil and water conservation.

Based on the context presented, we propose the design of a tokenization platform following the methodology proposed in the previous section.

4.1 Ecosystem Value Proposition

What problem does the proposed solution address for the ecosystem as a whole?

The proposed solution addresses the problem of the lack of timely and quality information needed to develop effective climate asset monetization models. It facilitates the creation of these models, enabling early-stage trading of climate assets, and provides robust and accessible green accounting for both the public and private sectors. This initiative not only drives environmental sustainability through the tokenization of carbon credits, but also promotes tree planting as a key player in this process, thus generating the inherent value of carbon sequestration in forests.

How would the blockchain-based solution affect each of the participants currently involved and what value propositions will the participants in the ecosystem be able to agree on?

The implementation of blockchain would allow landowners and investors to place their tokens in more sophisticated markets, obtaining a higher value than they could obtain in traditional local markets. This technology offers transparency, security, and efficiency in the trading of climate assets, which benefits both landowners by increasing the value of their carbon credits, and investors by providing access to more diversified and profitable investment opportunities. It also encourages active participation in climate change mitigation by facilitating the financing of reforestation and conservation projects, thus strengthening the global commitment to environmental sustainability.

4.2 Identification of Participants

Who are the participants and why would they want to join the ecosystem?

Landowners: The blockchain provides a secure and trans-parent platform for registering land ownership, protecting landowner rights, and ensuring the legitimacy of transactions. It facilitates landowner participation in sustainable forest management and environmental conservation practices, such as certifications and carbon programs. It also allows for the tokenization of forest assets, facilitating decentralized investments and the generation of income through environmental services and carbon credits, thus promoting forest conservation.

Investors: By stimulating the development of an early market for carbon offsets, the technology enables the establishment of efficient financing mechanisms at a global level, ensuring short, medium, and long-term sustainability of projects through the exchange of carbon tokens, regardless of the maturity stage of the climate project.

Local and national government: Governments facilitate collaboration between different stakeholders, including landowners, local communities, non-governmental organizations, and the private sector. In the context of the blockchain, this is essential to develop shared standards, interoperable platforms, and information systems that are accepted and used by all stakeholders involved in forest management and the fight against deforestation. Local governments are responsible for implementing public policies and incentive programs that encourage forest conservation and sustainable management. The blockchain can facilitate the implementation of incentive mechanisms such as payments for environmental services, carbon offset programs, and conservation funds, ensuring that these resources reach those who are actually contributing to forest protection.

Why do participants consider the new value proposition and solution desirable?

One of the great advances of blockchain is the ability to tokenize the protocol or infrastructure layer of information technology on the Internet. This capability allows a new economic model to emerge in which open source, cooperative, and distributed innovation is possible and necessary to maintain and increase the value of the token that represents access to the underlying infrastructure. By tokenizing the common asset, economically generative applications can be built, facilitating an attractive asset for both developers and funders.

In addition, blockchain tokenization offers a level of transparency and security, eliminating intermediaries and reducing associated costs. This not only strengthens trust among participants in the carbon bond ecosystem, but also increases the liquidity and accessibility of climate assets. Thus, landowners and investors can benefit from more efficient and fairer markets, where the value of carbon credits accurately reflects their ecological and economic impact, thus driving continued commitment to global sustainability.

Will the ecosystem be open or closed?

The described ecosystem is an open system. In an open ecosystem, diverse actors, such as buyers and sellers, are allowed to freely participate and trade tokens at an early stage. This approach initiates a dynamic and competitive carbon removal market, incentivizing innovation and adaptation to changing market conditions.

The creation of an early carbon removal market indicates a flexible and evolving environment. This market encourages practices such as reforestation and other forms of carbon sequestration, facilitating the entry of new participants and promoting competition. In contrast, a closed ecosystem would be more restrictive, limiting participation and reducing opportunities for innovation and competition in carbon sequestration.

How does the blockchain technology available today enable the envisioned value propositions?

Blockchain technology not only improves transparency and efficiency in forest asset management, but also offers innovative tools to combat deforestation and promote sustainable practices, integrating economic, environmental, and social aspects in a more equitable and effective way.

- It enables the creation of transparent and verifiable records of every transaction related to forest assets, from the purchase and sale of land to the harvesting and transport of timber. This helps prevent illegal logging and provides a reliable chain of custody from the source to the end consumer.

- Through smart contracts, it is possible to automate compliance with forestry regulations and standards, such as sustainable management certificates (FSC, PEFC). This facilitates auditing and ensures that forestry practices are responsible and sustainable.
- Forest assets can be tokenized on the blockchain, enabling fractional ownership and decentralized investment in forestry projects. This can encourage conservation by making economic benefits more widely distributed and encouraging more people to participate in forest protection.
- Using Internet of Things (IoT) sensors and immutable records on the blockchain, the health of forest ecosystems can be monitored in real time. This includes air quality, biodiversity, and deforestation patterns, providing accurate data for environmental management and rapid response to illegal activities.
- Blockchain-based platforms allow direct financing of forestry projects through crowdfunding or investments, eliminating intermediaries and reducing costs. This can encourage reforestation and the conservation of protected areas.
- Through mechanisms such as carbon tokens, local communities and landowners can receive compensation for keeping their forests standing and storing carbon, which encourages sustainable practices and reduces pressure on forest resources.

Who are the participants needed to be successful?

- Landowners: They are crucial as they own the land where reforestation projects and other carbon sequestration initiatives can be implemented.
- Investors: They provide the capital necessary to finance carbon removal projects, ensuring their long-term viability and sustainability.
- Local and national government: Play a vital role in creating policies and regulations that encourage and facilitate carbon sequestration initiatives. In addition, they can provide incentives and technical support.
- Communities: The participation and support of local communities is critical to the success of projects. Their involvement ensures that projects are sustainable and aligned with local needs and priorities.

What are the incentives for each of the required participants to engage in the ecosystem, and how does the blockchain solution shape a viable business future for them?

- Landowners and investors: Revenues based on the monetization of the business model.
- Local and national government: New business models to monetize the local economy.
- Communities: Protection of forest assets, participation in new business models.

4.3 Ecosystem Model

What are the possible models to implement?

Federated: Most access is through a web interface made available to the user by the administrators.

How might the model change over time?

As carbon tokens mature and become more valuable, the carbon removal market model may undergo significant changes. As tokens mature, their market value will increase, attracting more investors and participants, which will strengthen the market and promote more carbon removal projects. Transparency and traceability will be enhanced through public reporting of token transactions, allowing all stakeholders, including project owners, traditional investors, and token owners, to verify the authenticity and tracking of climate assets in real time. In addition, the use of advanced technologies such as blockchain to record and verify transactions will become more common, improving security and facilitating project tracking and auditing. Regulations and standards will also evolve, with governments and regulators developing stricter regulations to ensure market integrity. Increased participation and diversification of actors, including large corporations and investment funds, will contribute to an increase in the quantity and quality of carbon removal initiatives. Finally, the increased availability of information and the professionalization of the market will lead to improvements in operational efficiency, making verification and certification processes faster and less costly. In short, these changes will result in a more transparent, reliable, and efficient carbon market, benefiting all participants and promoting the sustainable growth of carbon removal projects.

4.4 Governance Model

What consortium governance will be needed to bring it to life?

The development of protocols will be required for the verification of climate outcomes. In addition, continued engagement and support for emerging movements around regenerative agriculture, ecological stewardship of land, water, and air, and alternative economics will be necessary. The token will enable betting on validators to secure the network, access to trusted data that feeds end-user protocols and applications, and will be used to reward climate outcomes. The tokens will run natively on the platform layers.

What governance model will constitute a fair, scalable and robust arrangement for all?

The tokens constitute debt backing and allow financiers to quickly access the immutable information of climate projects, making it easy to access resources from funds such as simple or syndicated debt, investment funds, grants or subsidies, and green or sustainable bonds.

When the project is banked or is at the stage of raising financing, the transactions of partial financing advances are shared on the blockchain, generating confidence for more investors or financiers to support the totality of the projects.

4.5 Preparation

What are the resources and capabilities needed for the initial development of the ecosystem?

- Traceability system supported by IoT for the capture and management of information related to soils, crops, and premises.
- Satellite data and images
- Primary information from surveys and open government data.
- Human capital for the development of technological applications based on IoT, data analysis, and artificial intelligence.
- Human capital for socialization and training in topics related to blockchain and new business models.

Is the proposed solution legal?

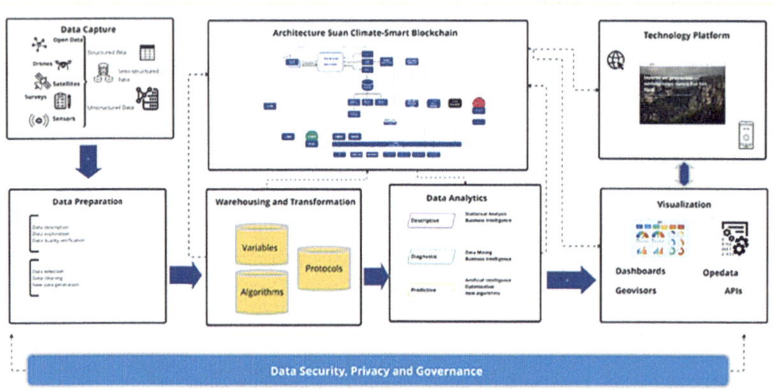

Fig. 2. IT Architecture for Blockchain

There is no regulatory framework in Colombia that regulates cryptocurrencies, however, it is in the process of being regulated by means of a bill. The National Tax and Customs Directorate accepted the existence of cryptoassets and classifies them as intangible assets (intangibles). In this sense, the Banco de la República has stated that cryptocurrencies do not obey any regulation since they are not any type of officially recognized financial instrument for the realization of transactions.

What are the main risks and possible errors?

- Community rejection due to lack of knowledge or distrust of the system.
- Risk of changes in blockchain regulation or environmental regulation.
- Availability of quality data and information.
- Transactions are not covered by any type of private or state guarantee.

4.6 IT Architecture Proposal

The following is the proposed 1.0 architecture for the traceability of forest assets. The architecture is proposed from four major blocks: the first is oriented to data enlistment through data capture, understanding, and preparation; the second in data storage and analytics; the third block includes the entire architecture based on blockchain along with its technology platform; and the fourth is the block of information deployment through visualizations as shown in Fig. 2.

5 Conclusion

Blockchain technology can be used to help reduce deforestation in a number of ways. One potential use case is to create a decentralized carbon market where transactions for buying and selling carbon credits would be recorded on a blockchain. Smart contracts would be used to automatically execute the transfer of ownership of credits which would facilitate tracking the flow of carbon credits, and ensure that they are used for their intended purpose. In addition, the use of smart contracts could automate the process of compliance and reporting to regulators, allow consumers and businesses to make more informed decisions, and promote sustainable sourcing practices.

An MVE for blockchain includes all the necessary components to quickly lay the foundation for a thriving ecosystem, that can grow and evolve over time by connecting users, suppliers, partners, and developers. It allows companies to quickly launch and test their ecosystem idea. Stakeholders interact creating a sense of ownership and loyalty to improve the ecosystem and guide development. Challenges for the project that need to be carefully considered and addressed are that all stakeholders must be aligned and incentivized to contribute to the ecosystem, and that it can be vulnerable to security breaches and other cyber threats such as data privacy breaches and intellectual property theft.

Future work will include the development of the tokenization platform that supports the ecosystem and the collection of information from various sources such as IoT, satellites, and primary information to establish the inventory of forest assets in the area.

References

1. Sunny, J., Undralla, N., Madhusudanan Pillai, V.: Supply chain transparency through blockchain-based traceability: an overview with demonstration. Comput. Ind. Eng. **150**, 106895 (2020). https://doi.org/10.1016/j.cie.2020.106895. https://www.sciencedirect.com/science/article/pii/S0360835220305829
2. Jiménez Valbuena, J.A., Osorio Sanabria, M.A., Maestre Góngora, G.P.: Análisis del estado de adopción de la tecnología blockchain en el sector financiero colombiano. Revista CIES **12**(3) (2021). http://revista.escolme.edu.co/index.php/cies/article/view/372
3. World Economic Forum: Redesigning trust: Blockchain deployment toolkit - ecosystem (2020). https://widgets.weforum.org/blockchain-toolkit/. Accessed 23 Dec 2024

4. Gawer, A., Cusumano, M.A.: Industry platforms and ecosystem innovation. J. Prod. Innov. Manag. **31**(3), 417–433 (2014)
5. Adner, R.: Ecosystem as structure: an actionable construct for strategy. J. Manag. **43**(1), 39–58 (2017)
6. Mahmudnia, D., Arashpour, M., Bai, Y., Feng, H.: Drones and blockchain integration to manage forest fires in remote regions. Drones **6**(11) (2022)https://doi.org/10.3390/drones6110331. https://www.mdpi.com/2504-446X/6/11/331
7. Kotsialou, G., Kuralbayeva, K., Laing, T.: Blockchain's potential in forest offsets, the voluntary carbon markets and redd+. Environ. Conserv. **49**(3), 137–145 (2022)
8. Nandhini, J., Anuratha, K., Sangeetha, K., Jaswant, K.A.: Smart tree management with biodiversity preservation using image processing and blockchain technology. In: 2021 International Conference on System, Computation, Automation and Networking (ICSCAN), pp. 1–6 (2021)https://doi.org/10.1109/ICSCAN53069.2021.9526540
9. Shah, S., Jansen, S.: A framework for designing self-sustaining ecosystems with blockchain. In: Klotins, E., Wnuk, K. (eds.) ICSOB 2020. LNBIP, vol. 407, pp. 184–192. Springer, Cham (2021). https://doi.org/10.1007/978-3-030-67292-8_14
10. Abdon Almeida, L., de Souza, C., Magalhães Lima, A., Quites Reis, R.: A case study on the usage of the value blueprint for ecosystem design. In: Proceedings of the XI Brazilian Symposium on Information Systems (SBSI 2015), p. 43 (2015). https://aisel.aisnet.org/sbis2015/43
11. Papanikolaou, E., Angelis, J., Moustakis, V.: Which type of ecosystem for distributed ledger technology? Technol. Soc. **72**, 102143 (2023). https://doi.org/10.1016/j.techsoc.2022.102143. https://www.sciencedirect.com/science/article/pii/S0160791X22002846
12. Adner, R.: Winning the Right Game: How to Disrupt, Defend, and Deliver in a Changing World. The MIT Press, Cambridge (2021)
13. Boyd, S.: Minimum viable ecosystem (2023). https://medium.com/work-futures/minimum-viable-ecosystem-53ae03d43cbf. Accessed 12 Jan 2023
14. Živkovic, P., McCurdy, D., Zou, M., Raymond, A.H.: Mind the gap: tech-based dispute resolutions in global supply blockchains. Bus. Horiz. **66**(1), 13–26 (2023)
15. Muma, S.: Your blockchain project is doomed if you don't ask these five questions. Supply Chain Management Review (2020). https://www.scmr.com/article/your_blockchain_project_is_doomed_if_you_dont_ask_these_five_questions. Accessed 23 Dec 2023
16. Corporación Autónoma Regional de la Orinoquia: Plan de acción 2020–2023 (2020). https://corporinoquia.gov.co/es/la-corporacion/informacion-general/planes-de-accion-corporinoquia/plan-de-accion-2020-2023.html. Último acceso: 23 de diciembre de 2024

Optimization of Data Storage and Transmission in Wearable Devices for ADLs and Falls Using Bluetooth Low Energy: An Approach to Minimize Packet Loss

Manny Villa[1,2(✉)] and Eduardo Casilari[1]

[1] Departamento de Tecnología Electrónica, Instituto TELMA, Universidad de Málaga, Málaga, Spain
{mannyvilla,ecasilari}@uma.es

[2] Programa de Ingeniería Electrónica, Universidad de Investigación y Desarrollo (UDI), Bucaramanga, Colombia
mvilla2@udi.edu.co

Abstract. This paper describes a practical experience with the real-time collection and storage of data from a wearable device used to generate essential datasets for the development of algorithms and artificial intelligence models. The aims of the models are to identify activities of daily living (ADL) movements and to detect falls. The device is equipped with accelerometer, gyroscope, magnetometer, barometer, and microphone sensors, all operating at a frequency of 20 Hz. Initially, significant packet loss was observed during real-time transmission due to distance and data saturation transmitted by Bluetooth Low Energy (BLE). To address this issue, an algorithm was developed to store data in a buffer for 10 s of movement. Subsequently, the data is transmitted to the receiving mobile device in a controlled manner when it is near the wearable device, achieving packet loss-free data transfer. The packet size was specified to analyze its impact on transmission. Two mobile applications were used: one for real-time transmission, and another for collecting data stored in the buffer. Both applications allow data to be saved in .csv files for subsequent visualization. During buffer transmission, a transmission frequency of 10 Hz was implemented on the wearable device to ensure no data loss. This device is designed for data collection in a laboratory setting to obtain raw data that will allow future researchers to create models for identifying ADL movements and falls.

Keywords: Data Storage Optimization · Wearable Devices · Bluetooth Low Energy (BLE) · Packet Loss Minimization · Real-Time Data Collection

1 Introduction

The proliferation of wearable devices has driven the development of technologies that enable real-time data collection and transmission for various applications,

F. Suni-Lopez et al. (Eds.): CIIS 2024, ISEM 48, pp. 203–215, 2025.
https://doi.org/10.1007/978-3-031-91481-2_14

ranging from health monitoring to tracking sports performance [1]. The ability of these devices to collect data from multiple sensors and transmit it is crucial for their effectiveness and utility in practical applications, such as the generation of new datasets [2].

Bluetooth Low Energy (BLE) is a wireless communication technology that has been widely adopted in wearable devices due to its low power consumption and its ability to transmit data efficiently. It allows for low-power communication, which is ideal for devices that need to operate for long periods with small batteries [3].

However, continuous high-frequency data transmission can lead to significant issues, such as packet loss. This loss can be caused by several factors, including the distance between the device and the receiver, environmental interference, and the saturation of the communication channel due to the high density of transmitted data. Packet loss can result in incomplete or inaccurate data, negatively affecting the effectiveness of the wearable device. Various studies have proposed strategies to mitigate packet loss in BLE systems. Some of these strategies include retransmission techniques and error-correcting codes to improve data integrity [4,5]. However, these solutions can increase power consumption and latency, which is not ideal for wearable devices that rely on long-lasting batteries.

In this study, we present an alternative solution that involves the temporary storage of data in a buffer, followed by controlled transmission. This approach aims to ensure the integrity of the collected data and improve the reliability of transmission in wearable devices while minimizing power consumption. Two mobile applications were used: one for real-time transmission, and another for collecting data stored in the buffer. Both applications allow data to be saved in .csv files for subsequent visualization. During buffer transmission, an adjusted transmission frequency was implemented in the Arduino to ensure no data loss. It is worth noting that this device is specifically designed for data collection in a laboratory setting to obtain raw data that will enable future researchers to create models for identifying ADL movements and falls.

2 Related Work

Bluetooth Low Energy has emerged as a pivotal technology in the advancement of wearable devices, particularly for health monitoring, smart home solutions, and diverse IoT applications, thanks to its low-power communication efficiency. This section reviews recent advances in BLE technology, focusing on strategies to improve reliability of data transmission and reduce packet loss in BLE's connectionless modes, which are critical for long-term monitoring without frequent maintenance.

Studies conducted by Liu et al. [6] detailed the evolution of BLE, highlighting its reduction in power consumption and increase in transmission range, which are beneficial for IoT applications. They explored modulation and coding techniques at the physical layer, noting improvements in transmission latency and power consumption, crucial for wearable devices that operate continuously over long periods.

Tsimbalo et al. [5] addressed the challenges of packet loss in BLE's connectionless mode by proposing a combination of techniques, including the use of orthogonally polarized antennas, cyclic redundancy check error correction, application layer coding, and the implementation of multiple spatially distributed receivers. Their novel approach significantly improved data transmission reliability in residential environments, crucial for healthcare applications where data integrity and reliability are paramount.

Narasimhan [7] developed a fall detection system using a skin-contact sensor equipped with a triaxial accelerometer and BLE to continuously transmit acceleration data to a computer. The sensor, positioned on the user's chest, measures accelerations at a frequency of 125 Hz, allowing for accurate fall detection. The experiments highlighted the advantages of using BLE, particularly its low energy consumption, which enables prolonged device operation. This approach also facilitates the execution of a threshold-based algorithm externally on a computer, utilizing the continuous data transmission for detailed and precise analysis.

Yu et al. [8] implemented a fall detection system using BLE to transmit alarm messages through a network of relay nodes. The experiments demonstrated that the nodes could organize hierarchically and transmit messages effectively, allowing the user's location to be determined by the received signal strength indicator (RSSI). However, challenges such as higher energy consumption in continuous transmission mode and message duplication were identified, highlighting the need to optimize energy efficiency and reliability in sensor networks for fall detection.

Additionally, Tipparaju et al. [4] made significant contributions to the study of packet loss mitigation in wearable healthcare ecosystems using BLE. They proposed a queued packet transmission protocol that allows for flexible, real-time, or accumulated data transmission, adapting to the end user's needs. This protocol has been shown to reduce packet loss to less than 1% in systems based on both Android and iOS, providing a robust solution for critical health applications where reliable data transmission is essential.

These studies collectively highlight progressive improvements in BLE technology, particularly in enhancing the reliability and efficiency of data transmission for wearable technologies. Ongoing developments ensure that BLE remains a viable option for future applications requiring reliable long-term operation without intensive maintenance.

In our study, we have taken these advances a step further by integrating a multi-sensor system into a wearable device that uses BLE not only for real-time transmission, but also for efficient data storage in a buffer. This allows for more adaptable and robust data collection, crucial for accurate monitoring in medical and health applications.

3 Materials and Methods

3.1 System Description

The system consists of a wearable device equipped with multiple sensors, a BLE communication module, and two mobile applications for data collection and visualization. The main components of the system are as follows:

- **Sensors:** The device includes an accelerometer, gyroscope, magnetometer, barometer, and microphone, all operating at a sampling frequency of 20 Hz.
- **BLE module:** Used for data transmission to the mobile device.
- **Mobile applications:** Two applications developed in Flutter; one for real-time data transmission and another for collecting data stored in the buffer.
- **Wearable device:** The device is placed on the chest, specifically at the sternum level.

3.2 Wearable Device Configuration

The wearable device primarily consists of the Arduino Nano 33 BLE Sense Rev 2, which integrates several advanced sensors for detailed data collection on movements and the environment. This device is placed on the chest, specifically at the sternum level, as shown in Fig. 3 [9]. The integrated sensors are as fol-lows:

- **BMI270 6-axis IMU**
  - 3-axis accelerometer with ranges of ±2 g, ±4 g, ±8 g, and ±16 g.
  - 3-axis gyroscope with ranges of ±125 dps, ±250 dps, ±500 dps, ±1000 dps, and ±2000 dps.
- **BMM150 3-axis IMU (magnetometer)**
  - Digital 3-axis geomagnetic sensor with a resolution of 0.3 µT.
  - Measurement ranges of ±1300 µT on the x and y axes, and ±2500 µT on the z axis.
- **LPS22HB barometer**
  - Measures pressures from 260 to 1260 hPa with a 24-bit resolution.
  - Built-in temperature compensation and a data output range from 1 Hz to 75 Hz.
- **MP34DT06JTR digital microphone**
  - Sensitivity of –26 dBFS ±3 dB and a signal-to-noise ratio of 64 dB.

Power Supply. The device is powered by a 2000 mAh lithium battery (Easy-lander 103450 model), which provides sufficient energy to operate the sen-sors and the BLE module during data collection sessions.

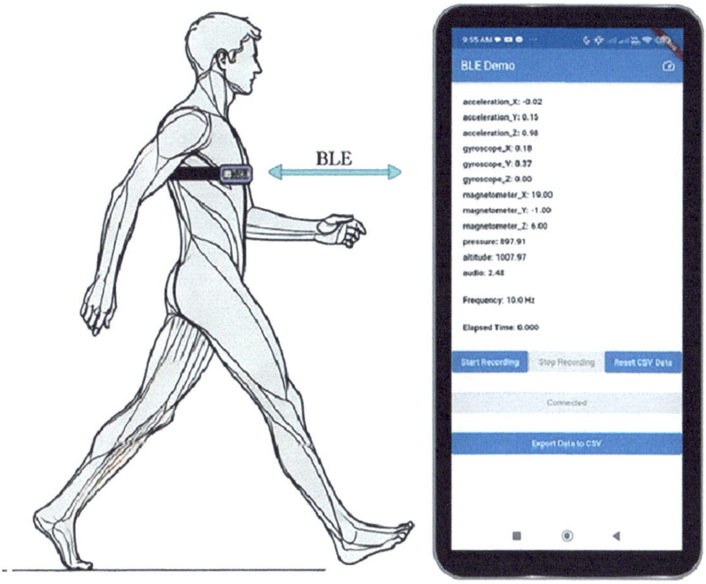

Fig. 1. System architecture for real-time data capture.

3.3 Real-Time Data Transmission

Figure 1 illustrates the initial system configuration, where sensor data is transmitted in real-time to the mobile device via BLE. In this setup, each sensor generates data at a frequency of 20 Hz, which is immediately sent to the mobile device.

Transmission Process. Figure 2 shows the flowchart of the real-time data trans-mission process from the wearable device. This process begins with the initializa-tion of the sensors and the BLE module. Once initialized, the system continuous-ly reads data from the sensors. If the BLE central device is connected, the data is immediately sent to the mobile device. If the connection to the BLE central de-vice is lost, the system attempts to reconnect. This reconnection loop continues until the BLE central device is reconnected. During transmission, it is continu-ously checked whether the data transmission should be stopped. If not, the system continues to send data; otherwise, the transmission process ends.

Mobile Application for Real-Time Transmission. The mobile application, developed in Flutter and shown in Fig. 1, is designed to receive and display data transmitted in real-time from the wearable device via BLE. This application is responsible for visualizing the sensor data as it is received, providing instant and continuous visualization. Additionally, the application allows the user to stop data reception at any time, and then, by pressing a button on the mobile app, the received data can be downloaded into a .csv file for subsequent analysis. These data remain in the mobile's memory until the reset command is executed

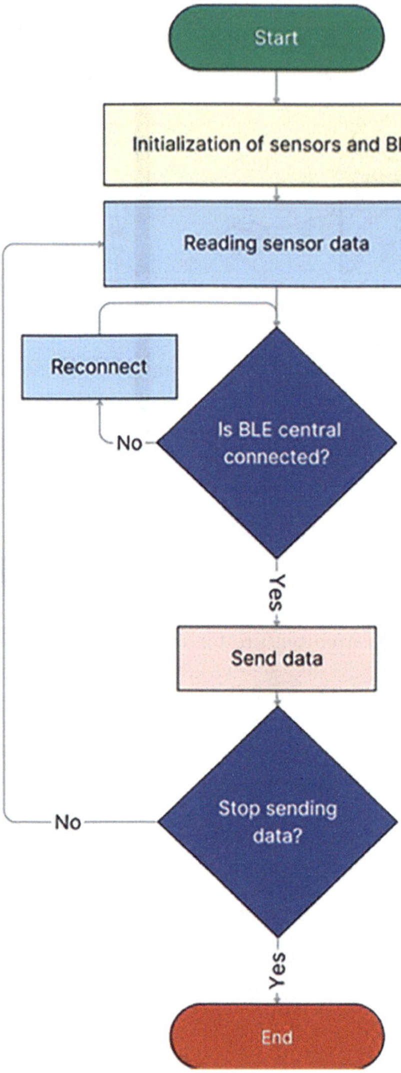

Fig. 2. Real-time data transmission flowchart

for a new capture. This facilitates real-time data management and allows for a detailed review of the captured movements.

3.4 Buffer Storage and Controlled Transmission

Figure 3 illustrates the enhanced system configuration, where data is temporarily stored in a buffer for 10 s of movement at a frequency of 20 Hz. Subsequently, the stored data is transmitted to the mobile device in a controlled manner when it is near the device and transmission is initiated.

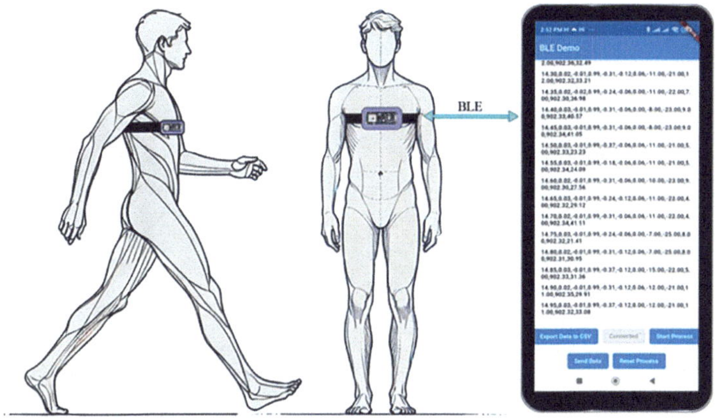

Fig. 3. System architecture for data capture using buffer storage and controlled transmission

Storage and Transmission Process. Figure 4 shows the flowchart of the data storage and transmission process of the wearable device. The process begins with the initialization of the sensors and the BLE module. Sensor data is read and stored in an internal buffer for 10-s intervals. Then, it checks if the BLE central device is connected. If not connected, the system continuously attempts to reconnect. If connected, the system decides whether to initiate data transmission from the buffer. If transmission is to be initiated, the data is sent to the mobile device with an adjusted frequency and optimized packet size to avoid oversaturation of the BLE connection. The process ends once all data has been transmitted.

Mobile Application for Buffer Transmission. The mobile application shown in Fig. 3 is designed to receive data that has been previously stored in the buffer of the wearable device. After 10 s of data storage, the user can initiate transmission from the buffer using the application. Data is transmitted in controlled packets to avoid oversaturation of the BLE connection. Once all data is received, the application saves the information in a .csv file, allowing for detailed subsequent analysis. This application ensures the integrity and order of the data collected during the storage interval, facilitating the export and analysis of the raw data.

3.5 Packet Content and Export to .CSV File

In both transmission cases (real-time and from the buffer), sensor data is packaged along with a timestamp that identifies the exact moment each reading was taken. Each packet contains the following sensor readings: acceleration (X, Y, Z axes), gyroscope (X, Y, Z axes), magnetometer (X, Y, Z axes), pressure, altitude, and audio level. These data are exported to a .csv file, where each row represents

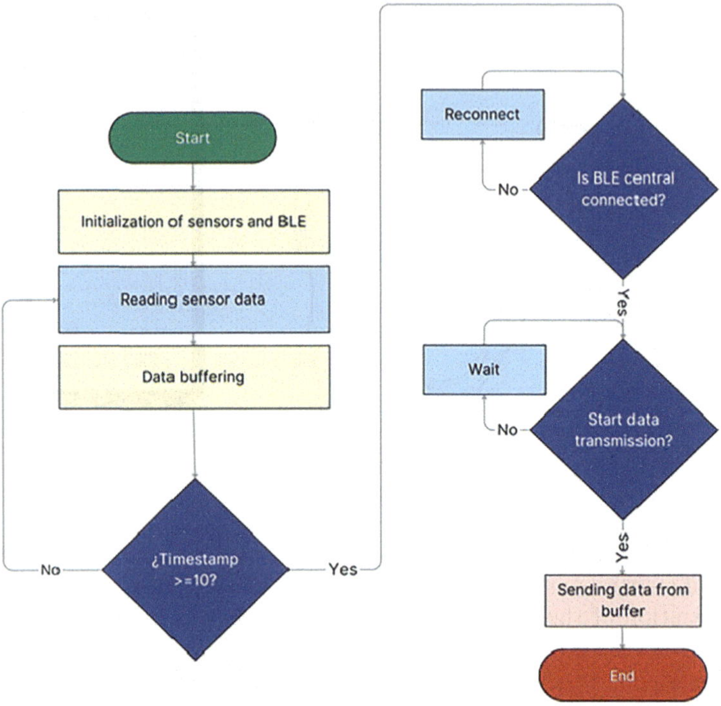

Fig. 4. Controlled data transmission flowchart

a sensor reading, facilitating the subsequent analysis of the collected data. This file structure allows researchers to use data analysis tools to study movements and detect falls accurately and efficiently. Figure 5 shows how the data exported to a .csv file is organized from the wearable device.

	A	B	C	D	E	F	G	H	I	J	K	L
1	Timestamp	Acce_x	Acce_y	Acce_z	Gyro_x	Gyro_y	Gyro_z	Magn_x	Magn_y	Magn_z	Pressure	Audio_level
2	0.00	-0.02	0.90	0.41	-2.38	1.10	-1.59	-1.00	-27.00	-4.00	899.23	34.50
3	0.05	-0.02	0.90	0.41	-2.01	-2.01	0.61	-1.00	-27.00	-4.00	899.30	16.85
4	0.10	-0.02	0.90	0.42	0.43	0.43	-0.37	-1.00	-28.00	-6.00	899.22	15.04
5	0.15	-0.03	0.89	0.43	0.67	0.92	0.79	-1.00	-28.00	-6.00	899.24	27.81
6	0.20	-0.03	0.88	0.42	2.32	1.28	-2.08	1.00	-27.00	-7.00	899.26	12.61
7	0.25	-0.02	0.89	0.42	-0.31	-0.37	-0.49	1.00	-27.00	-7.00	899.26	21.31
8	0.30	-0.02	0.89	0.42	-0.85	0.24	-1.83	-2.00	-30.00	-5.00	899.24	27.46
9	0.35	-0.01	0.90	0.41	-2.87	-2.14	-0.61	-3.00	-30.00	-2.00	899.26	21.73
10	0.40	-0.01	0.88	0.41	2.08	1.16	-0.92	-3.00	-30.00	-2.00	899.24	17.86
11	0.45	-0.01	0.87	0.43	0.61	0.43	-1.40	1.00	-27.00	-4.00	899.26	19.05
12	0.50	-0.01	0.89	0.42	-0.79	0.00	-1.95	1.00	-27.00	-4.00	899.25	13.83
	...	...	...	...	...	...	...	...	...	...	...	...

Fig. 5. Representation of data in the .csv file

4 Testing

To validate the system's functionality, tests were conducted with four daily activities (ADLs) and four types of falls, each within a 10-s period and repeated four times, allowing for the collection of a detailed dataset. The ADLs included walking forward, jumping, climbing stairs, and bending down and standing up. The types of falls evaluated were forward, backward, right side, and left side falls.

Activities and falls were performed ensuring that the device was positioned at the sternum level. Using the device's sensors, which include an accelerometer, gyroscope, magnetometer, barometer, and digital microphone, data was collected at a sampling frequency of 20 Hz. This data was transmitted in real-time via BLE to the mobile application, where it was visualized in real-time. Additionally, the application allowed data reception to be stopped at any time to save the data in a .csv file.

Subsequently, the same movement and fall tests were performed using the second method of buffer storage and controlled transmission. In this method, data was stored in the device's buffer for 10 s before being transmitted to the mobile device. Once storage was completed, data transmission was initiated from the mobile application, ensuring efficient and lossless transmission. The received data was also exported to .csv files for subsequent analysis.

Figure 6 graphically represents the data obtained from a forward fall, showing acceleration, gyroscope, and magnetometer data on all three axes. This visualization allows for the observation of specific patterns captured by the sensors during the fall event, providing a useful tool for detailed analysis and validation of the fall detection system, as well as for other movements.

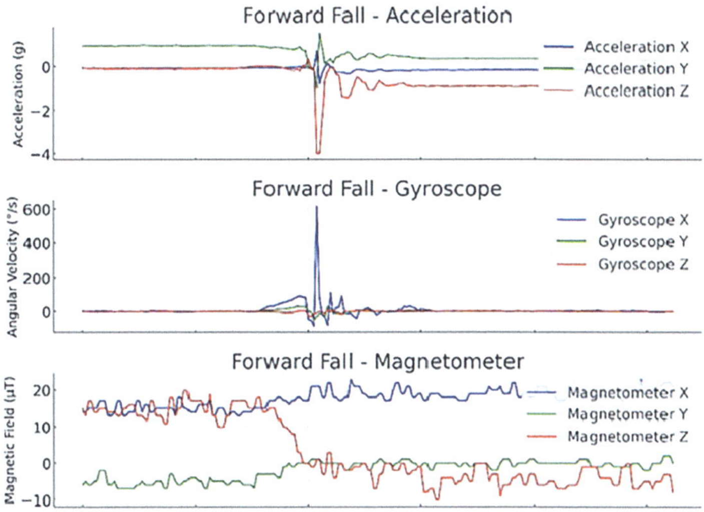

Fig. 6. Graphical representation of the forward fall

5 Results

To assess the effectiveness of the data transmission methods, tests were conducted using two different approaches: real-time transmission, and buffer storage with controlled transmission. Various ADLs and different types of falls were tested. The activities evaluated included walking forward and climbing stairs, both performed over a distance of 5 m, as well as falls in different directions (forward, backward, right lateral, and left lateral), jumping, and bending and standing up, all of which were conducted from a stationary position. The results focus on the absence of expected timestamps as an indicator of data loss, considering the lack of an expected timestamp as a packet loss. Packet loss analysis was performed by calculating the time intervals between each data sample using the timestamps. These intervals were compared to the expected sampling interval (e.g., 50 ms), and any significantly larger intervals were recorded as potential data losses. Subsequently, the number of anomalous intervals was summarized as a percentage of the total intervals. For each activity, 200 packets were considered, corresponding to a 10-s period, from timestamp 0.00 to 9.95. Any absence of an expected timestamp within this interval was considered a packet loss. The findings are displayed in Table 1.

Table 1. Packet Loss Analysis by Activity and Transmission Method

Category	Activity Type	Lost Packets	Packet Loss (%)
Falls	Forward	3	1.5
Falls	Backward	2	1.0
Falls	Right Lateral	4	2.0
Falls	Left Lateral	1	0.5
ADLs	Walking Forward	0	0.0
ADLs	Jumping	5	2.5
ADLs	Climbing Stairs	2	1.0
ADLs	Bending and Standing Up	1	0.5

The packet loss analysis using real-time transmission, conducted at a distance of 1.5 m, shows that the loss varies between 0% and 2.5%. The highest packet loss was observed during the "Jumping" activity, with a loss of 2.5%. Right side falls exhibited a packet loss of 2.0%, the highest among the fall categories. No packet loss was observed in the "Walking forward" activity.

Packet Loss Analysis Using Buffer Storage and Controlled Transmission. Using the buffer storage and controlled transmission method, conducted at a distance of 1 cm-which is an appropriate distance to ensure data retrieval after generation and storage, guaranteeing transmission without packet loss due to distance-and with a reduced transmission speed of 10 Hz, no packet loss was observed in any case within the experimental setting. It is worth noting that this approach is well-suited for generating datasets that will later be used to train artificial intelligence models for fall detection. This indicates that the buffer storage and controlled transmission method is more robust and effective in preventing data loss during the collection and transmission of sensor data. This analysis demonstrates the superiority of the buffer storage and controlled transmission method over real-time transmission, especially in scenarios where data loss could compromise the accuracy and reliability of the movement and fall detection system.

6 Discussion

This article discusses the optimization of data collection and storage from wearable devices, focusing on the development of datasets for the analysis ofADLs and fall detection. By reviewing relevant literature and integrating an innovative method of buffer storage and controlled transmission via BLE, this study offers concrete solutions to common challenges in real-time data transmission, such as packet loss due to data saturation and distance.

The work by Liu et al. [6], which explores the efficiency of BLE in reducing power consumption and improving transmission, provides important technical context that supports the development of the proposed method in this study. Additionally, the research by Tsimbalo et al. [5] on the implementation of orthogonally polarized antennas and cyclic redundancy check error correction is fundamental to understanding how to enhance data reception and integrity.

The proposed method of buffer storage and controlled transmission has significant implications for the design and operation of wearable devices in health monitoring and fall detection. By improving the quality of collected data and ensuring efficient operation in real-world environments, the method is crucial for applications that require continuous and non-intrusive monitoring. This is vital for the viability of wearable health technologies in managing chronic conditions and monitoring patients.

Moreover, the accuracy and integrity of the collected data are fundamental for training artificial intelligence models that can detect and predict critical events with high reliability. The quality of the dataset directly influences the effectiveness of algorithms developed to identify patterns in ADLs and falls. Future research could benefit from exploring the robustness of this buffer storage and controlled transmission method in various network configurations and with different data loads. Evaluating the adaptability of the method across different technological platforms and its interaction with other emerging technologies could provide a broader understanding of its applicability and limitations. This approach not only enhances the accuracy of data collected for critical applications, but also establishes a solid foundation for the continuous development of advanced technologies in the health field and beyond.

7 Conclusion

This study demonstrated the effectiveness of buffer storage and controlled transmission in mitigating packet loss in wearable devices using BLE. Two data transmission methods were compared: real-time transmission, and buffer storage followed by controlled transmission. Real-time transmission showed significant packet loss, especially in high-intensity activities such as jumping (2.5%) and side falls (2.0%). These results confirm that continuous high-frequency transmission can saturate the communication channel, increasing the likelihood of packet loss due to interference and distance between the device and the mobile receiver.

Buffer storage and controlled transmission proved to be highly effective in eliminating packet loss. By storing data for 10 s and transmitting it at a frequency of 10 Hz, data integrity was maintained without oversaturating the BLE connection. This approach aligns with recommended mitigation strategies in the literature, such as reducing transmission frequency and using retransmission protocols and error correction codes to improve data integrity. Implementing a buffer for temporary data storage followed by controlled transmission is particularly useful in critical applications where data integrity is essential, such as fall detection and health monitoring. This method not only ensures the accuracy of the collected data, but also optimizes power consumption-a crucial factor for wearable devices operating on long-lasting batteries.

The methodology used facilitates the creation of robust datasets, essential for developing algorithms and artificial intelligence models aimed at identifying ADL movements and detecting falls. The integrity and accuracy of the data are crucial for training highly reliable artificial intelligence models.

In conclusion, adopting buffer storage and controlled transmission represents a robust solution for data collection and transmission in wearable devices. This approach overcomes the limitations observed in real-time transmission and offers greater reliability in data integrity, which is crucial for health monitoring and fall detection applications. Additionally, it establishes a strong basis for the creation of advanced algorithms and artificial intelligence models.

References

1. Rathi, N., Kakani, M., El-Sharkawy, M., Rizkalla, M.: Wearable low power pre-fall detection system with iot and bluetooth capabilities. In: 2017 IEEE National Aerospace and Electronics Conference (NAECON), pp. 241–244 (2017). https://doi.org/10.1109/NAECON.2017.8268778
2. Tosi, J., Taffoni, F., Santacatterina, M., Sannino, R., Formica, D.: Performance evaluation of bluetooth low energy: a systematic review. Sensors **17**(12) (2017). https://doi.org/10.3390/s17122898. https://www.mdpi.com/1424-8220/17/12/2898
3. Gomez, C., Oller, J., Paradells, J.: Overview and evaluation of bluetooth low energy: an emerging low-power wireless technology. Sensors **12**(9), 11734–11753 (2012)

4. Tipparaju, V.V., Mallires, K.R., Wang, D., Tsow, F., Xian, X.: Mitigation of data packet loss in bluetooth low energy-based wearable healthcare ecosystem. Biosensors **11**(10) (2021).https://doi.org/10.3390/bios11100350. https://www.mdpi.com/2079-6374/11/10/350
5. Tsimbalo, E., et al.: Mitigating packet loss in connectionless bluetooth low energy. In: 2015 IEEE 2nd World Forum on Internet of Things (WF-IoT), pp. 291–296 (2015). https://doi.org/10.1109/WF-IoT.2015.7389068
6. Liu, C., Zhang, Y., Zhou, H.: A comprehensive study of bluetooth low energy. J. Phys. Conf. Ser. **2093**(1), 012021 (2021). https://doi.org/10.1088/1742-6596/2093/1/012021
7. Narasimhan, R.: Skin-contact sensor for automatic fall detection. In: 2012 Annual International Conference of the IEEE Engineering in Medicine and Biology Society, pp. 4038–4041 (2012). https://doi.org/10.1109/EMBC.2012.6346853
8. Yu, T., Stamm, A., Hartanto, R.: Design and implementation of a bluetooth low energy-based local area network for fall detection. Proceedings **2**(6) (2018). https://doi.org/10.3390/proceedings2060289. https://www.mdpi.com/2504-3900/2/6/289
9. Arduino: Arduino nano 33 ble sense rev2 documentation (2024). https://docs.arduino.cc/hardware/nano-33-ble-sense-rev2/. Accessed 3 Sept 2024

Social and Cultural Impact

Socioeconomic and Environmental Factors Influencing Women's Entry Into STEM Careers: A Systematic Review

Marilia Fernanda Jurado Vilca[1(✉)], Jadeolinda Fernandez Castillo[1], Franci Suni-Lopez[2], Guillermo Antonio Dávila[1], and Nadia Rodríguez-Rodríguez[1]

[1] Universidad de Lima, Avenida Javier Prado Este 4600, Santiago de Surco, Lima, Peru
{20203953,20200775}@aloe.ulima.edu.pe, {gdavila,nrodrigu}@ulima.edu.pe

[2] Laboratorio de Inteligencia Artificial, Universidad de Lima, Avenida Javier Prado Este 4600, Santiago de Surco, Lima, Peru
fsuni@ulima.edu.pe

Abstract. Historically and globally, there has been a notable under-representation of women in STEM (science, technology, engineering, and mathematics) careers. This disparity is influenced by various factors, including gender stereotypes and women's self-perception of their academic abilities. The present systematic review aims to explore the application of machine learning techniques to predict women's entry into STEM careers, with a focus on identifying and analyzing socioeconomic and environmental factors that influence their decision-making. The review was conducted using academic repositories such as Scopus and Web of Science, resulting in the selection of 40 articles. These studies included educational records, surveys conducted with students, preprocessing techniques, machine learning algorithms, and metrics for information analysis. Key findings revealed that the support vector machine (SVM) model was the most commonly used, while random forest (RF) and gradient boosting (GB) algorithms demonstrated superior performance. Regarding validation metrics, the majority of studies utilized accuracy, precision, F1 score, recall, and ROC curve to assess model effectiveness. In conclusion, the present review provides a comprehensive understanding of the gender gap in STEM fields, highlighting the significant influence of specific factors and offering insights into predictive modeling for women's participation in these areas.

Keywords: STEM · women · gender gap · systematic review

1 Introduction

The STEM acronym, originally from the United States of America, groups together the disciplines of science, technology, engineering, and mathematics.

F. Suni-Lopez et al. (Eds.): CIIS 2024, ISEM 48, pp. 219–232, 2025.
https://doi.org/10.1007/978-3-031-91481-2_15

The main objective of education within STEM disciplines is to foster cognitive skills in students, such as problem-solving, ingenuity, and critical thinking, with the purpose of preparing them to face current challenges and enhance societal progress with the help of science [34].

Throughout history, significant female roles in STEM have been recognized, taking as example Ada Lovelace as the first programmer; Marie Curie as the only person to win two Nobel Prizes from different disciplines; and Katherine Johnson, Dorothy Vaughan, and Mary Jackson as the women who led the United States of America to victory in the space race during the Cold War, among others. At present, the numbers and statistics are alarming and reflect a worrying situation. According to the information presented by the United Nations Educational, Scientific, and Cultural Organization (UNESCO), the global representation of women in STEM careers reaches only 35% [30]. In the Peruvian context, the situation is not very different, as figures show that only 32% of graduates in STEM careers are women, while only 46% of Peruvian women aspire to enter a STEM career, compared to 78% of men [31]. Given this, there is an urgent need to comprehensively address the existing gender disparity in careers involved in STEM fields, in order to encourage interest and participation among women.

The existence of a gender gap in STEM careers is not a trivial matter due to the long-term impact it has on both the personal and professional trajectories of students. To this day, there is a decline in participation where the marginalization of women in STEM-related careers and jobs harms their independence and capacity for empowerment [6]. Consequently, the absence of female representation in STEM-related careers consolidates gender stereotypes, decreasing their interest in entering these areas [15]. From another perspective, the gap also leads to a lack of diversity of thoughts, which limits the multiplicity of visions and perceptions of problems, reducing the ability to address challenges that hinder innovation and progress [5].

There are a large number of previous studies that demonstrate how, through machine learning algorithms, it is possible to predict the entrance of women into STEM careers and the importance of the factors that condition their decisions. Some of the studies have focused on the students' academic performance and school life, and how these aspects shape their scientific identity [1,3,18,34]; some on the student's performance in basic mathematics courses, which open doors to interest in STEM fields [17,19]; and others on the scores of international knowledge tests such as PISA, INVALSI, or SIMCE [10,23,26,29].

The present study seeks to conduct a literature review on the use of machine learning to predict the entry of women into STEM careers, in order to identify and examine the socio-economic and environmental factors that influence their decision. In this way, the present study is relevant at a social level, and also for its practical implications. From a social perspective, the aim is to determine the main factors that marginalize women in STEM careers, since the low percentage of women enrolled in higher education in these areas creates a disproportion in terms of employment, health, and educational opportunities. As for the practical implications, by using machine learning algorithms as the main tool for diagnosis, it is expected not only to determine patterns that contribute to the accentuation of discrimination and bias, but also to propose concrete and measurable solutions that encourage female participation in STEM.

2 Methodology

Currently, machine learning is a concept widely used in different subject areas such as medicine, politics, and social sciences. The great availability of data sources and computing capacity, algorithms, and calculation efficiency are key pieces for a prediction model in quantitative research [8]. In the context of our study, by applying machine learning, it is possible to anticipate the entry of women into STEM careers by identifying and understanding the factors that discourage them. Therefore, in order to thoroughly understand the relationship between STEM education and gender, an exhaustive search was carried out in databases including Scopus and Web of Science, using terms such as 'STEM education', 'STEM career', 'machine learning', and 'gender'. These searches yielded a large number of relevant articles, of which 40 were selected. In order to elaborate a systematic review of the literature, the following research questions are presented, which will give way to a categorized discussion in the subsequent section:

RQ1. What are the main factors that influence women's decision to choose STEM careers?

RQ2. What data sets are available to predict women's participation in STEM careers?

RQ3. What machine learning algorithms are used to predict women's choice of a STEM career?

RQ4. What metrics are used to evaluate the performance of predictive models in predicting women's choice of STEM careers?

For this, the population, intervention, comparison, outcome, and context (PICOC) methodology was applied, which focuses on offering a structured approach that allows the formulation of questions in a systematic way, component by component [21]. Population refers to the people or individuals grouped together to be studied. Intervention refers to the treatment or action to be investigated. Comparison evaluates the effects of the intervention. The result measures the effectiveness of the intervention; and the context indicates the environment or context of the study. In this way, the following scenario is presented:

Population: Students of high school or general studies courses at universities
Intervention: Socio-economic and environmental factors
Comparison: Predictive analysis models
Outcome: Efficiency indicators of predictive models
Context: Entry into STEM fields

3 Results

3.1 Factors that Influence Women's Decision to Opt for STEM Careers

Over the years, the number of women attending university has increased; however, educational inequalities between women and men in the choice of professional careers are still evident. Recent research indicates that gender differences in the choice of higher education courses persist. Women tend not to select careers related to STEM. This preference is influenced by a variety of factors, including social and economic context, as well as family and school characteristics [23].

The choice for STEM careers is usually influenced by various factors; therefore, in this section, student records will be analyzed to identify the factors that hinder their entry into these fields. Therefore, it is crucial to identify the relevant variables to determine the inclination of women towards a STEM career. Following the proposal of [25], it is suggested to categorize each associated factor into three categories, based on its importance: positive, which refers to findings that show a beneficial or favorable effect on the choice of STEM careers; negative, which indicates findings that show an adverse or harmful effect; and without significance, which describes findings that do not demonstrate a significant relationship between certain factors and the choice of STEM careers.

Various studies [2,3,17,18] have pointed out that the gender factor positively influences the entry into STEM careers. However, despite the positive results mentioned, some studies such as that of [23] do not find a statistical significance in the influence of gender, suggesting that other factors could moderate this impact.

On the other hand, academic factors of the students such as participation in STEM-related subjects, the level of mathematical proficiency, the grade point average (GPA), and the final average in STEM-related courses, in several studies, have been shown to have a positive influence [1,2,17,26]. In contrast, academic performance, participation in extracurricular subjects, and average grades in different courses have a negative influence [18].

Factors such as nationality, ethnicity, and geographical region can have both negative and positive significance, depending on the context [3,7,9,17,18,23]. These factors imply that the geographical and cultural context can play a crucial role in STEM education, which suggests the need to adapt educational programs to local realities to ensure their effectiveness.

Finally, in relation to the variables that do not show statistical significance, various studies have investigated factors such as parental education, disability, and time spent playing video games [7,18]. The absence of statistical significance in relation to these factors could indicate that their influence is indirect or that they interact in a complex way with other factors, not being reflected in a direct analysis.

In summary, several factors influence the choice of STEM careers, such as family support, gender, and attitudes towards science, as well as adverse factors

Table 1. Main influencing factors in entering STEM careers

Author(s)	Factors	Significance		Not significant
		Positive	Negative	
[28]	Number of courses	*		
	Interest in robotics/statistics	*		
	Selection of high-paying careers	*		
	Participation in extracurricular subjects		*	
	Academic performance		*	
	Gender			*
	Age			*
[2]	Gender	*		
	Participation in subjects related to the STEM field	*		
	Social Pressure		*	
	Cumulative average grade in secondary education			*
	Cumulative grade point average of the first year			*
	Cumulative grade point average for the last year			*
[9]	Gender	*		
	Family influence	*		
	Socioeconomic status		*	
	Education level of father and mother			*
	Ethnicity			*
	Disability			*
	Interest in science/mathematics/technology			*
[19]	Attitude towards chemistry	*		
	Autonomy towards chemistry	*		
	Age			*
	Gender			*
	Participation in laboratories			*
[7]	Aspiration to STEM careers	*		
	STEM motivation	*		
	STEM courses	*		
	STEM credits	*		
	Performance in mathematics	*		
	Demographics		*	
	Parental involvement		*	
	School location		*	
	Extracurricular programs			*
[18]	Gender	*		
	Ethnicity	*		
	Geographical region	*		
	Mathematics level	*		
	Satisfaction with classes	*		
	Weighted average		*	
	Socioeconomic status		*	
	Hours spent on video games per day			*

(continued)

Table 1. (*continued*)

Author(s)	Factors	Significance		Not significant
		Positive	Negative	
[3]	Study modality	*		
	Ethnicity	*		
	Age	*		
	Gender	*		
	GPA	*		
	Course notes			*
[26]	Weighted average	*		
	Course attendance	*		
	Score in the mathematics test	*		
	Language test score	*		
	Socioeconomic status		*	
	Rural school		*	
	Public school		*	
	Parent education			*
[1]	Type of school	*		
	Final average in science courses	*		
	Weighted average	*		
	Gender			*
	Demographics			*
[17]	Admission score (APS)	*		
	Academic score in mathematics/science/physics	*		
	Ethnicity		*	
	Nationality		*	
	Gender			*
[23]	INVALSI score in mathematics	*		
	INVALSI score in Italian	*		
	Socioeconomic status of the school		*	
	Gender			*
	Region			*

such as socioeconomic level and social pressure. In addition, there are variables whose effect is not statistically significant. In order to carry out an accurate analysis in predicting the choice of STEM careers by women, it is important to consider factors with both positive and negative impact (see Table 1).

3.2 Datasets

Public datasets stand out for their accessibility in the research community, being often used by more than one author. Given what has been presented and according to what can be seen in Table 2, the most used dataset was The ASSISTance Longitudinal dataset with 1,709 gross records [4,16,32,33]. However, this dataset has some drawbacks such as the limited number of records and the age of the data, since it collects information on students from the years 2004-2007-20 years ago. The different time frames could represent the biggest problem, as it is likely that the information does not reflect the current reality. For this reason, the authors who used this dataset agree that, for future studies, more extensive and temporally relevant datasets should be utilized.

The largest dataset in terms of records was the PISA 2015 database, which had 519,334 raw records and was used by [29]. A disadvantage of this dataset is that it tends to produce bias due to a large number of countries participating

Table 2. Main public datasets

Dataset used	Year of collection	Number of records	Variables	Subject of study
Recommendation System for Graduate Level Universities in the United States dataset	2015	53,645	26 variables grouped into: – Academic	Applicants at 45 different universities in the United States
Link:	https://github.com/joemanley201/universityRecommendationSystem/			**Referenced by:** [27]
High School Longitudinal Study of 2009 (HSLS: 09)	2017	25,210	102 variables grouped in: -Demographic -Relatives -Academic -Socio-economic	Students from the first grade in high school up to their post-secondary studies (11th grade, 12th grade, three and four years after high school)
Link:	https://nces.ed.gov/surveys/hsls09/hsls09_data.asp			**Referenced by:**[7, 18]
Open University Learning Analysis dataset (OULAD)	2017	32,593	43 variables grouped in: -Demographic -Academic	Students of different E-Learning courses related to STEM
Link:	https://analyse.kmi.open.ac.uk/open_dataset			**Referenced by:** [20, 22, 34]
The ASSISTment Longitudinal dataset	2017	1,709	78 variables grouped in: -Demographic -Academic -Psychological	High school students who used the ASSISTment learning platform during the years 2004–2007
Link:	https://sites.google.com/view/assistmentsdatamining/data-mining-competition-2017?authuser=0			**Referenced by:** [4, 16, 32, 33]
PISA 2015 database	2017	519,334	25 variables grouped in: -Demographic -Relatives -Academic	High school students from 72 different countries who participated in the PISA test in the year 2015
Link:	https://www.oecd.org/pisa/data/2015database/			**Referenced by:** [29]
PISA 2018 Spain	2019	35,943	25 variables grouped in: -Demographic -Relatives -Academic	Spanish high school students who participated in the PISA test that was held in 2018
Link:	https://www.oecd-ilibrary.org/sites/22e49155-en/index.html?itemId=/content/component/22e49155-en			**Referenced by:** [10]
Records obtained from the National Institute for the Evaluation of the Education and Training System (INVALSI) of Italy.	2019	Not specified	7 variables grouped in: – Demographic -Socio-economic -Academic	13th grade students from Italian schools who participated in the INVALSI test
Link:	https://serviziostatistico.invalsi.it/en/invalsi_ss_data/microdati-campione-g13-2018-19/			**Referenced by:** [23]

in the PISA test, being that the countries do not reflect the same reality, which can be demonstrated by the low resulting metrics.

The High School Longitudinal Study of 2009 (HSLS:09) has the largest number of variables, 102. This dataset was used in the studies of [7,18], where both agreed that the dataset presents a great advantage in terms of its longitudinal approach. This is due to it not only focusing on collecting information from first grade high school students, but also tracks them in the following years of high school, as well as three and four years after high school. However, a slight disadvantage was anticipated in this regard, as it was difficult to keep track of all the students, leading to loss of records during the process.

Finally, other relevant datasets have been found which deserve to be mentioned. Among them, there is PISA 2018 Spain, which like the PISA 2015 database, comes from the PISA test, but is focused on a single country to reduce the potential bias that could occur. Also, there is the Open University Learning Analysis dataset, which was used in a significant number of studies because it collected information on the academic development of general studies students who had taken STEM courses virtually. Finally, there is the Recommendation System for Graduate Level Universities in the United States dataset, which is a dataset intended for recommending courses based entirely on the academic performance of high school students in the United States.

3.3 Validation Algorithms and Metrics

This section seeks to explore and analyze the different algorithmic approaches that have been used in recent research to address the problem raised. Additionally, the evaluation metrics used to measure the effectiveness and accuracy of these predictive models were discussed. This analysis will not only provide a comprehensive overview of the current state of knowledge in the field, but will also establish a solid framework for the implementation of effective strategies in the proposed research.

For this section, 12 relevant studies have been selected (see Table 3), which denote diversity due to the wide range of countries they contain and the number of records used, which varies from 325 [13] to 53,645 [27]. This highlights the importance of testing multiple algorithms for each specific problem, since the algorithm's performance varies according to the dataset. Among the algorithms that showed higher systematic performance, there are support vector machines (SVMs), neural networks, and ensemble methods such as random forest (RF) and gradient boosting (GB). While these models tend to be complex, some simple models such as decision trees seek to show competitive results.

The studies selected in Table 3 use a wide set of validation metrics and it is notable that some stand out more than others for their ability to differ between the perspectives presented by the model. In particular, accuracy, recall, and F-1 score metrics are key indicators to correctly identify true positives. In this way, the aforementioned complex models are interpreted as those that offer a better result beyond what the accuracy metric may indicate. An example of this is clearly noticeable in the study by [11], where the recall metric of the SVM

Table 3. Analysis of metrics of different algorithms

Dataset Author(s)	Best Algorithms	Acc.	Prec.	Recall	F1-score	ROC curve	Others
Texas A&M University-Commerce student survey - No. of records not mentioned [28]	Support Vector Machines	0.985	0.985	0.985	0.985	–	–
Students at educational institutions in Bangladesh -364 records [12]	Random Forest	0.849	0.849	0.846	0.843	–	–
Students from 45 different universities in the United States - 53,645 records [27]	Ensemble Neural Network	0.910	0.910	0.910	0.910	–	–
High school students from Benin City-325 records [13]	Decision Tree	0.970	–	0.960	0.980	–	–
	Random Forest	0.990	–	0.980	0.990	–	–
	Linear SVC	0.970	–	0.980	0.980	–	–
Students of the faculties of engineering at Andhra Pradesh National University of Technology (India) - 1027 records [24]	XGBoost	0.860	0.840	–	0.860	0.770	TP Rate = 0.870 and FP Rate = 0.150
High school students in Istanbul, Turkey - 600 registrations [11]	Support Vector Machines	0.887	0.887	1,000	0.940	0.500	TP Rate = 1
	K-STAR	0.928	0.933	0.991	0.961	0.926	TP Rate = 0.991
	J48	0.887	0.887	1,000	0.940	0.487	TP Rate = 1
Queensborough Community College of CUNY - 23,000 records [3]	K-Nearest Neighbors	0.842	0.866	0.947	0.905	–	Jaccard Score = 0.826
	Decision Tree	0.849	0.874	0.945	0.908	–	Jaccard Score = 0.832
Students of a high school in Konya (Turkey) - 600 records [35]	Support Vector Machines	0.948	0.936	0.948	0.941	0.990	Kappa = 0.915
Saudi Arabia Ministry of Education - 35,406 registrations [1]	Decision Tree	0.970	0.970	0.970	0.970	0.984	–
	Random Forest	0.980	0.980	0.980	0.980	0.999	–
High school students from two Portuguese schools - 395 records [14]	Naíve Bayes	0.912	0.917	0.912	0.913	–	–
Organization for Economic Cooperation and Development - 35,943 registrations [10]	Gradient Boosting	0.865	0.669	0.863	0.865	0.934	Spec = 0.546
	Support Vector Machines	0.863	0.668	0.871	0.861	0.937	Spec = 0.542
University of Johannesburg in South Africa - 29,158 registrations [17]	K-Nearest Neighbors	0.965	0.965	0.965	0.964	–	RSME = 0.554
	Decision Tree	0.967	0.968	0.967	0.967	–	RSME = 0.525
	Random Forest	0.966	0.966	0.966	0.966	–	RSME = 0.534

and J48 algorithms reaches 100%, while both accuracies denote 88.7%, which is low when compared to the accuracies of the K-STAR, RF, or random tree algorithms from the same study. Likewise, studies by [1,27], using the mentioned metrics demonstrate a high index for the RF algorithms and neural networks, respectively.

Finally, Table 3 highlights the importance of choosing an appropriate model or algorithm depending on the nature of the dataset, such as the number of records or variables and the objectives of the study. A complex model could suffer from over-fitting on the training data, while a simple model might fail to capture the underlying relationships between variables, sacrificing accuracy.

4 Discussion

Throughout the present analysis, significant influence of aspects such as gender, academic performance, participation in extracurricular activities, family influence, nationality, and socioeconomic level has been highlighted in the decisions of women regarding their professional life. Although the factors did not have the same degree of impact or level of significance, they are part of the complexity that women face when choosing these fields. Various studies have shown that women face gender stereotypes that affect their self-perception and self-confidence in their scientific and mathematical abilities, a crucial factor during high school for considering a STEM career [17].

The vast majority of findings agree that female representation in STEM careers is not simply an individual problem, but a partial combination of family, academic, and socioeconomic influences. For example, the influence of gender and performance in STEM subjects during high school are factors identified as critical determinants in vocational decision-making [2,18]. In addition, the literature review shows that gender has a positive impact in certain contexts; that is, its significance varies according to geographical region and cultural context. This suggests that interventions must be adapted to local realities in order to be effective [7,10].

Regarding the influence of the family environment, studies have shown that family support and expectations of academic success have a significant impact on women's decision to pursue a STEM career [9], as an individual's confidence begins within the family. On the other hand, women from families with high socioeconomic levels are more likely to receive the necessary encouragement to excel in STEM disciplines, while the lack of resources and support in less advantaged contexts can limit access to quality educational opportunities [23].

In the geographical and cultural contexts, factors such as ethnicity and geographical region can influence both positively and negatively [7,18]. For example, in some countries or regions, women may be more exposed to educational programs that promote gender equity in STEM; while in others, gender stereotypes may be more entrenched, limiting opportunities and interest.

The importance of these factors has been the result of a prominent application of machine learning algorithms, specifically those for classification and prediction of complex behavior. In this way, SVM was, at the same time, one of the most used and most efficient algorithms, reaching significantly-high metrics, as has been shown in the study by [11] with a 100% recall. However, its application may be limited by the computational complexity it requires, especially in large data sets. Likewise, ensemble algorithms such as RF and GB have determined a robust performance when correlating variables. The RF algorithm, after SVM, has been reported to have the best metrics as in the study by [1], where a recall of 98% is reported; GB, following the study by [29], demonstrates an accuracy of 86.5%.

5 Conclusions

The present study conducted a literature review on the use of machine learning to predict the entry of women into STEM careers, in order to identify and examine the socio-economic and environmental factors that influenced their decision. The literature indicates, first of all, that gender, academic performance in science and mathematics, family influence, nationality, and socioeconomic level are some of the influential factors as to why a woman does not choose to study a STEM career.

Regarding datasets, a great variety stood out due to their nature, since some contained a good number of records, while others a large number of variables. Likewise, the origin of the data is noteworthy, as it includes a diversity of countries, which implies cultural, social, and economic differences that, as observed, can influence the final outcome when selecting the most relevant factors that determine whether a woman chooses a STEM career or not.

As for the machine learning algorithms, the most used and most efficient were SVM, RF, GB, and neural networks. These models, mostly complex, were selected to make accurate predictions on large datasets, significantly contributing to the robustness and reliability of predictive analyses.

The metrics considered in the literature are quite extensive; however, the most commonly used were accuracy, precision, recall, F-1 score, and the ROC curve, which have proven to be fundamental for the performance of each model. In particular, RF, XGBoost, and neural networks have offered high and balanced performance across all metrics, highlighting their validity in making predictions.

Finally, future research could conduct predictive studies in developing or emerging countries due to their limited participation in the reviewed literature. Additionally, a comparative analysis could be conducted between influencing factors in countries with similar cultural contexts in order to focus the research based on that factor. Likewise, the number of variables that could have more relevance in a more current context such as the influence of social networks or the impact of the media could be expanded.

References

1. Abdalkareem, M., Min-Allah, N.: Explainable models for predicting academic pathways for high school students in Saudi Arabia. IEEE Access **12**, 30604–30626 (2024). https://doi.org/10.1109/ACCESS.2024.3369586
2. Adekitan, A., Shobayo, O.: Gender-based comparison of students' academic performance using regression models. Eng. Appl. Sci. Res. **47**(3), 241–248 (2020). https://doi.org/10.14456/easr.2020.27, https://ph01.tci-thaijo.org/index.php/easr/article/view/218107

3. Akcay Ozkan, Z., Yuanhong, Y., Stelmach, E.: Predicting student success in college algebra classes using machine learning. In: Proceedings of the 2023 ASEE Annual Conference & Exposition. ASEE Conferences, Baltimore, Maryland (2023). https://doi.org/10.18260/1-2-43933, https://peer.asee.org/43933
4. Almeda, M.V., Baker, R.S.: Predicting student participation in stem careers: The role of affect and engagement during middle school. J. Educ. Data Min. **12**(2) (2020). https://doi.org/10.5281/zenodo.4008054, https://par.nsf.gov/biblio/10190395. Country unknown/Code not available
5. Barreal Rodríguez, A.A.: Reducir la brecha de gnero en disciplinas stem: propuesta de actuacin en matemáticas (2021). https://dspace.uib.es/xmlui/bitstream/handle/11201/157723/tfm_2020-21_MFPR_abr544_4237.pdf?sequence=1&isAllowed=y. Accedido: 2024-12-27
6. Cabero-Almenara, J., Valencia-Ortiz, R.: Stem y gnero: un asunto no resuelto. Revista De Investigacin Y Evaluacin Educativa **8**(1), 4–17 (2021). https://doi.org/10.47554/revie2021.8.86
7. Chang, C.N., Lin, S., Kwok, O.M., Saw, G.K.: Predicting STEM major choice: a machine learning classification and regression tree approach. J. STEM Educ. Res. **6**(2), 358–374 (2023)
8. Chen, Y., Wu, X., Hu, A., He, G., Ju, G.: Social prediction: a new research paradigm based on machine learning. J. Chin. Soc. **8**(1), 1–21 (2021). https://doi.org/10.1186/s40711-021-00152-z
9. Garca-Ramos, L.E., Nieto Bernal, W., Zambrano Vizcaino, Y.P., Osorio Salcedo, V., Barbosa Gutierrez, V.M.: Desarrollo de una aplicacin web para la orientacin vocacional y promocin de carreras STEM implementando tcnicas de data mining. Proyectos finales Pregrado en Ingeniera de Sistemas y Computacin (2021). http://hdl.handle.net/10584/9857. Publisher: Barranquilla, Universidad del Norte, 2021
10. Gmez-Talal, I., Bote-Curiel, L., Rojolvarez, J.L.: Understanding the disparities in mathematics performance: an interpretability-based examination. Eng. Appl. Artif. Intell. **133**, 108109 (2024). https://doi.org/10.1016/j.engappai.2024.108109. URL https://www.sciencedirect.com/science/article/pii/S0952197624002677
11. Gktepe Krpeoglu, S., Gktepe Yildiz, S.: Comparative analysis of algorithms with data mining methods for examining attitudes towards STEM fields. Educ. Inf. Technol. **28**(3), 2791–2826 (2023)
12. Hasan, R., Ovy, M.K.A., Nishi, I.Z., Hakim, M.A., Hafiz, R.: A decision support system of selecting groups (science/ business studies/ humanities) for secondary school students in bangladesh. In: 2020 11th International Conference on Computing, Communication and Networking Technologies (ICCCNT), pp. 1–6 (2020). https://doi.org/10.1109/ICCCNT49239.2020.9225411
13. Houngue, P., Hountondji, C., Dagba, T.: An effective decision-making support for student academic path selection using machine learning. Int. J. Adv. Comput. Sci. Appl. **13** (2022). https://doi.org/10.14569/IJACSA.2022.0131184
14. hui, Y.: Predicting math performance in high school students using machine learning techniques. Int. J. Adv. Comput. Sci. Appl. **15**(5) (2024). https://doi.org/10.14569/IJACSA.2024.0150516, http://dx.doi.org/10.14569/IJACSA.2024.0150516
15. Kenneth, A.: Gap in stem education: why is there a decline in women participation? Int. J. Res. STEM Educ. **4**, 55–63 (2022). https://doi.org/10.31098/ijrse.v4i1.369
16. Mandalapu, V., Gong, J.: Studying factors influencing feature on individual class prediction: the prediction of student stem and non-stem career choice. In: Desmarais, M., Lynch, C.F., Merceron, A., Nkambou, R. (eds.) Proceedings of the 12th International Conference on Educational Data Mining, pp. 607–610. International Educational Data Mining Society (2019)

17. Maphosa, M., Doorsamy, W., Paul, B.: Improving academic advising in engineering education with machine learning using a real-world dataset. Algorithms **17**, 85 (2024). https://doi.org/10.3390/a17020085
18. Nagy, M., Main, J., Molontay, R., Griffith, A.: Using machine learning methods to develop person-centered models predicting stem major choice. European Society for Engineering Education (SEFI) (2023). https://doi.org/10.21427/EE90-XX67
19. Ong, A.K.: A machine learning ensemble approach for predicting factors affecting stem students future intention to enroll in chemistry-related courses. Sustainability **14**, 16041 (2022). https://doi.org/10.3390/su142316041
20. Pelima, L., Sukmana, Y., Rosmansyah, Y.: Predicting university student graduation using academic performance and machine learning: a systematic literature review. IEEE Access **PP**, 1–1 (2024). https://doi.org/10.1109/ACCESS.2024.3361479
21. Petticrew, M., Roberts, H.: Systematic Reviews in the Social Sciences: A Practical Guide, vol. 11. Wiley (2006). https://doi.org/10.1002/9780470754887
22. Poudyal, S., Nagahi, M., Nagahi, M., Ghanbari, G.: Machine learning techniques for determining students' academic performance: a sustainable development case for engineering education. In: 2020 International Conference on Decision Aid Sciences and Application (DASA), pp. 920–924 (2020). https://doi.org/10.1109/DASA51403.2020.9317178
23. Priulla, A., Albano, A., D'Angelo, N., Attanasio, M.: A machine learning approach to predict university enrolment choices through students' high school background in Italy (2024). https://arxiv.org/abs/2403.13819
24. Rao, K.E., Pydi, B.M., Vital, T.P., Naidu, P.A., Prasann, U.D., Ravikumar, T.: An advanced machine learning approach for student placement prediction and analysis. Int. J. Performability Eng. **19**(8), 536 (2023). https://doi.org/10.23940/ijpe.23.08.p6.536546, https://www.ijpe-online.com/EN/abstract/article_4799.shtml
25. Roy-Garca, I., Rivas-Ruiz, R., Prez-Rodrguez, M., Palacios-Cruz, L.: Correlacin: No toda correlacin implica causalidad. Revista Alergia Mxico **66**(3), 354–360 (2019). https://doi.org/10.29262/ram.v66i3.651, https://revistaalergia.mx/ojs/index.php/ram/article/view/651
26. Sevilla, M., Luengo-Aravena, D., Faras, M.: Gender gap in stem pathways: the role of secondary curricula in a highly differentiated school system the case of chile. Int. J. STEM Educ. **10** (2023). https://doi.org/10.1186/s40594-023-00450-7
27. Sridhar, S., Mootha, S., Kolagati, S.: A university admission prediction system using stacked ensemble learning. In: 2020 Advanced Computing and Communication Technologies for High Performance Applications (ACCTHPA), pp. 162–167 (2020). https://doi.org/10.1109/ACCTHPA49271.2020.9213205
28. Suh, S., Upadhyaya, A., Nadig, A.: Analyzing personality traits and external factors for stem education awareness using machine learning. Int. J. Adv. Comput. Sci. Appl. **10** (2019). https://doi.org/10.14569/IJACSA.2019.0100501
29. Tan, L., Main, J., Darolia, R.: Using random forest analysis to identify student demographic and high school-level factors that predict college engineering major choice. J. Eng. Educ. **110** (2021). https://doi.org/10.1002/jee.20393
30. UNESCO: Nias, mujeres y stem: Cmo la fundacin ingeniosas ayuda a descubrir vocaciones en ciencias y tecnologa en chile y amrica latina (2023). https://www.unesco.org/es/articles/ninas-mujeres-y-stem-como-la-fundacion-ingeniosas-ayuda-descubrir-vocaciones-en-ciencias-y. Accessed 02 Jun 2024
31. UNICEF Per: #mschicastec: Nas y adolescentes mujeres en la ciencia y tecnologa. https://www.unicef.org/peru/ninas-adolescentes-mujeres-ciencia-tecnologia-stem. Accessed 02 Jun 2024

32. Yeung, C.K., Yeung, D.Y.: Incorporating features learned by an enhanced deep knowledge tracing model for STEM/non-STEM job prediction. Int. J. Artif. Intell. Educ. **29**(3), 317–341 (2019)
33. Zambrano, A., Baker, R.: Long-term prediction from topic-level knowledge and engagement in mathematics learning. In: Proceedings of the 14th Learning Analytics and Knowledge Conference, pp. 66–77 (2024). https://doi.org/10.1145/3636555.3636851
34. Zhang, J., et al.: E-learning behavior categories and influencing factors of stem courses: a case study of the open university learning analysis dataset (oulad). Sustainability **15**, 8235 (2023). https://doi.org/10.3390/su15108235
35. Çetinkaya, A., Baykan, O., Kirgiz, H.: Analysis of machine learning classification approaches for predicting students programming aptitude. Sustainability **15**, 12917 (2023). https://doi.org/10.3390/su151712917

Author Index

F. Suni-Lopez et al. (Eds.): CIIS 2024, ISEM 48, pp. 233–234, 2025.
https://doi.org/10.1007/978-3-031-91481-2

The manufacturer's authorised representative in the EU is Springer Nature Customer Service Centre GmbH, Europaplatz 3, 69115 Heidelberg, Germany. If you have any concerns regarding our products, please contact ProductSafety@springernature.com

Printed and bound by CPI Group (UK) Ltd, Croydon, CR0 4YY
30/09/2026
02225592-0010